NANOMETER MATERIALS THERMODYNAMICS: THE EFFECT OF RELATED PARAMETERS ON THE PROPERTIES OF NANOCRYSTALS AND ITS APPLICATIONS

纳米材料热力学：相关参数对纳米晶体性质的影响及应用研究

姜小宝 著

镇 江

图书在版编目(CIP)数据

纳米材料热力学：相关参数对纳米晶体性质的影响及应用研究 / 姜小宝著. — 镇江：江苏大学出版社，2018.12
 ISBN 978-7-5684-0996-4

Ⅰ. ①纳… Ⅱ. ①姜… Ⅲ. ①纳米材料－热力学 Ⅳ. ①TB383.13

中国版本图书馆 CIP 数据核字(2018)第 273280 号

纳米材料热力学：相关参数对纳米晶体性质的影响及应用研究
Nami Cailiao Relixue: Xiangguan Canshu Dui Nami Jingti Xingzhi De Yingxiang Ji Yingyong Yanjiu

著　　者	/ 姜小宝
责任编辑	/ 张小琴
出版发行	/ 江苏大学出版社
地　　址	/ 江苏省镇江市梦溪园巷 30 号(邮编：212003)
电　　话	/ 0511-84446464(传真)
网　　址	/ http://press.ujs.edu.cn
排　　版	/ 镇江市江东印刷有限责任公司
印　　刷	/ 句容市排印厂
开　　本	/ 890 mm×1 240 mm　1/32
印　　张	/ 4
字　　数	/ 113 千字
版　　次	/ 2018 年 12 月第 1 版　2018 年 12 月第 1 次印刷
书　　号	/ ISBN 978-7-5684-0996-4
定　　价	/ 25.00 元

如有印装质量问题请与本社营销部联系(电话：0511-84440882)

前言

近些年来,纳米尺度材料在现代技术应用和科学研究领域吸引了科学家们大量的兴趣。当材料尺寸减小到纳米尺度范围时,其光学、磁学、电学、催化及热力学性质都异于大块尺寸。纳米晶体的这些迷人的物理、化学性质主要是由其高的表面体积比,以及由此引起的表面原子或分子产生配位缺陷导致的。用理论方法描述纳米晶体性质变化和相应的相变过程是一项非常有意义的工作,不仅可以加深我们对纳米晶体相变本质的理解,而且可以引导实验操作制备,拓展它们在不同领域中的应用。

随着纳米科学和纳米技术的不断发展,相应的纳米材料热力学理论也得到了拓展。众多国内外学者根据不同的物理假设提出不同的热力学模型以描述纳米晶体的物理、化学及热力学性质,如熔化温度、能隙值、表面能、结合能、相变压力等随尺寸减小的变化趋势。然而,这些热力学理论都是以理想模型为基础建立的,即自由的纳米球体、圆柱纳米线及平面纳米薄膜。而在实际的制备和应用中,纳米晶体所处的环境和本身的结构往往偏离理想模型,比如不同的形状、成分、表面配位、压力等因素,都会对纳米晶体的性质产生重要影响。因此,有必要针对纳米晶体不同情况对相应的理论模型给予修正,以达到对纳米晶体性质的准确预测。

基于以上考虑,本书讨论了尺寸、形状、成分、维度、表面配位、压力等效应对纳米晶体性质变化的影响。本书共有9章:第1章为绪论,介绍纳米材料热力学的相关背景和研究意义;第2章讨论尺寸、形状、维度和成分对双金属纳米晶体的有序—无序转变温度和熔化温度的影响;第3章建立模型描述纳米微孔束缚的液晶材料的熔化与凝固转变过程,讨论界面条件对熔化和凝固转变温度的影响;第 4 章建模描述透明半导体 In_2O_3 纳米晶体由体心立方(bcc)向斜方六面体(rh)相变的过程,并着重分析表面能和表面应力对相变温度和相变压力的影响;第 5 章发展统一规则来评估固态相变的相变压力随尺寸减小的变化趋势,分析表面能、表面应力和原子单位体积对相变压力变化的影响;第 6 章讨论纳米晶体的尺寸和表面配位因素对表面能的影响,并将该模型分别应用于描述金属、半导体及合金体系;第 7 章通过建模,讨论尺寸和形状因素对 Fe_3O_4 纳米晶体 Verwey 转变温度的影响;第 8 章探讨尺寸、维度和压力效应对 ZnO 和 CdSe 纳米晶体能隙值的影响;第 9 章为结论与展望。

  本书在编写过程中参阅了大量的文献,在此向相关作者一并表示感谢。

目 录

第1章 绪论 001
  1.1 引言 001
  1.2 纳米材料的尺寸效应 002
  1.3 纳米材料的有序—无序转变 005
  1.4 纳米晶体材料熔化的尺寸效应 008
  1.5 固态相变的尺寸效应 013
    1.5.1 纳米固态相变热力学的研究现状 015
    1.5.2 热力学模型的建立基础 017
    1.5.3 纳米晶体表面应力对固态相变的影响 017

第2章 形状、尺寸和维度依赖的相变温度 019
  2.1 磁性二元合金的有序—无序转变温度 019
    2.1.1 磁性二元合金有序—无序转变背景简介 019
    2.1.2 二元合金有序—无序转变温度的尺寸与形状函数模型 020
    2.1.3 二元合金有序—无序转变模型和实验结果的对比与分析 022
    2.1.4 本章小结 025
  2.2 Au-Pt双金属纳米颗粒的熔化温度 025
    2.2.1 Au-Pt二元金属纳米晶体的熔化背景简介 025
    2.2.2 二元合金熔化温度的尺寸与形状函数模型的建立 026

2.2.3　Au-Pt 二元合金熔化温度理论模型和实验结果的
　　　　对比与分析　027
2.2.4　本章小结　030

第 3 章　液晶的熔化转变　031

3.1　液晶熔化的背景简介　031
3.2　液晶熔化的尺寸与界面效应函数理论模型的建立　032
3.3　理论模型与纳米微孔束缚的 PAA 和 5CB 液晶实验结果的
　　　对比与分析　035
3.4　本章小结　038

第 4 章　In_2O_3 纳米晶体相变热力学研究　039

4.1　In_2O_3 纳米晶体相变背景简介　039
4.2　In_2O_3 纳米晶体相变的温度与压力函数模型的建立　041
4.3　相变理论模型和实验结果的对比与分析　045
4.4　本章小结　049

第 5 章　确定影响纳米晶体相变压力变化趋势的因素　050

5.1　纳米晶体相变压力背景简介　050
5.2　确定纳米晶体相变压力变化趋势的理论方法　051
5.3　理论方法和实验结果的对比与分析　053
5.4　本章小结　057

第 6 章　尺寸和表面配位因素对纳米晶体表面能的影响　058

6.1　纳米晶体表面能背景简介　058
6.2　纳米晶体表面能的尺寸与表面配位函数模型的建立　059
6.3　纳米晶体表面能模型和金属、半导体、合金实验结果的
　　　对比与分析　061
6.4　本章小结　067

第 7 章 尺寸和形状因素对 Fe_3O_4 纳米晶体 Verwey 转变温度的影响 068

7.1 Fe_3O_4 纳米晶体的 Verwey 转变背景简介 068

7.2 Fe_3O_4 纳米晶体 Verwey 转变温度的尺寸与形状函数模型的建立 069

7.3 Fe_3O_4 纳米晶体 Verwey 转变温度理论模型与实验结果的对比与分析 071

7.4 本章小结 073

第 8 章 尺寸、维度和压力对 CdSe 和 ZnO 纳米晶体能带的调制效应 074

8.1 纳米晶体能带简介 074

8.2 纳米晶体能带的尺寸、维度和压力函数模型的建立 075

8.3 纳米晶体能带函数理论模型和 CdSe,ZnO 的实验结果的对比与分析 076

8.4 本章小结 079

第 9 章 总结与展望 081

参考文献 084

第 1 章 绪论

1.1 引言

宏观尺度(天体)和微观尺度(粒子)是现代科学发展的两个主流的研究方向。在微观尺度领域,纳米材料及纳米技术的研究逐渐成为现代科学研究的热点,更是与生物技术和信息技术并称为 21 世纪可以改变人类生存与发展方式的三大基础学科。在纳米尺度的微观领域,随着材料科学家们的不断探索,各种开拓性的研究工作相继展开,一个颠覆了传统概念,与大块世界截然不同的崭新的微观世界逐渐呈现在我们面前。纳米材料及纳米技术的持续发展将给人类的生活方式带来一场前所未有的革命。正如著名理论物理学家、Nobel 奖获得者 Richard P. Feynmann 教授在 1959 年预言的那样,"毫无疑问,当我们得以能够对细微尺度的事物加以操纵的话,将大大扩充我们可能获得的物性的范围。"

"Nano"一词起源于希腊语系,是侏儒的意思。它经常用作秒、米等单位的前缀,代表的是亿分之一的概念。纳米科学与纳米技术是一个涉及物理、化学、生物、工程、医疗等多领域的交叉学科。其作为现今国际上研究的热点,在不同的领域都带来了革命性的影响。纳米技术概念的提出,始源于 Richard P. Feynman 教授于 1959 年的一份题为 *There's plenty of room at the bottom* 的报告,他首次研究了在单个原子和分子尺度上改善材料性能的可行性,使得将大英百科全书写在一个针尖上成为可能。在这份报告中,他还预言了纳米尺度可以大大增强人们对物质的检测和控制能力。而

"纳米技术"一词的真正使用,是在1974年开始的。日本东京大学的年轻科学家Norio Taniguchi第一次用"纳米技术"一词描述在纳米尺寸上精确加工材料的能力。随着现代工业的发展,尤其是电子工业的发展,在芯片的制造领域对电子器件的尺寸及其制造工具的精确度提出了更高的要求。随着在技术领域中不断的应用,这一词汇在现代工业的各领域逐渐传播开来。

虽然纳米材料具有潜在的更加广阔的应用前景,但是对纳米材料的实用化进程还受制于其制备技术的发展与完善,以及人们对材料微观结构及性能的进一步深入的认识和理解。到现在为止,国际上对纳米材料的研究有很多,但是这些研究主要集中在纳米材料的制备技术,纳米材料不同于大块的特殊的化学、物理性能等领域,而对于纳米材料相变热力学等方面的研究工作还处于刚开始阶段,亟待进一步深入的探讨。美国科学院院士、热力学先驱Terrell L. Hill教授在1964年出版了 *Thermodynamics of Small systems* 一书,并且在2001年在 *Nano Lett* 杂志首卷指出:"然而,近年来随着在实验和理论上对纳米科学兴趣的激增,我想知道纳米热力学是否能够适用于平衡或亚平衡的纳米系统。"热力学在研究材料的亚稳性特征问题方面是一个非常实用的工具和手段,而随着尺寸的减小,纳米材料的结构特征与大块完整有序的固体结构有很大的偏离。因此,我们只能以现有的晶体相变热力学理论作为基础来研究纳米材料的热力学,针对其特点将相变热力学理论拓展到纳米尺度领域。

1.2 纳米材料的尺寸效应

介观是处于宏观和微观之间的一个过渡尺度范围,包括微米、亚微米、纳米直到团簇尺寸的范围。而纳米材料研究的范围是指尺寸为1~100 nm的粒子组成的粉末、薄膜、多层膜、大块、纤维等结构。由于其具有独特的结构特征(含有大量内界面),并表现出一系列的优异的物理、化学及力学性能,深入研究纳米材料不仅为进一步理解固体内部界面结构与性能提供了良好的契机,而且为提高材料的综合性能、发展新一代高性能材料创造了良好的条件。因此纳米材料迅速成为当前材料科学和凝聚态物理领域的研究热点。1981年,Gleiter教授首次提出"纳米晶体材料"(Nanocrystalline Materials)的设计理念。他把纳米晶体材料定义为材料的结构单元至少在一维方向上处于纳米量级的单相或多相材料。但是,当材料处于纳米尺度(或者界面尺寸至少有一维处于纳米尺度)的时候,其相应的材料性能及其表现出的性质不能够应用当前的经典物理学理论解释,而且,其所涉及的空间领域也超过了现有的实验技术所能够企及的范围。这就要求我们不得不对材料在宏观块体及微观分子、原子及亚原子两个极限状态下的物理、化学及电子性能有更深入的理解和掌握。2000年初,Gleiter再次对纳米材料(即广义的纳米晶体材料)进行了定义和分类。纳米材料既包括组成材料的结构单元,又包括材料自身尺度微观化(低维材料)的微结构的特征尺寸处于纳米量级的材料。根据这个定义,纳米材料大致可以分为三类:① 低维纳米材料,其中包括纳米粒子、纳米线、纳米缆、纳米管、纳米薄膜、纳米有机大分子等;② 表层纳米材料,即利用各种表面处理技术在纳米尺寸范围通过改变固体表面的化学成分或原子结构获得;③ 大块纳米材料,即以纳米量级的结构单元构成的块体材料。按照大块纳米材料的结构单元,其可以分为层状、柱状和球状三种类型。按照成分,它又可以分为:① 晶粒与界面成分相同(单相纳米材料);② 晶粒成分不同(为多相纳米材料);③ 晶粒和晶界成分不同(即某成分在界面处发生偏聚的纳米材料);④ 晶粒分布在不同化学和成分的基体中(例如析出纳米弥散相)。

  颗粒的尺寸是衡量纳米材料结构特性的重要结构参量之一,由于较高的界面密度是纳米材料的重要结构特征,材料的平均晶粒尺寸则可以直接反映界面密度含量。随着尺寸的减小,纳米材料具有与块体材料截然不同的光、电、磁、声、热力学等物理、化学及力学性能。例如,在光学性质上,光吸收增加并产生吸收峰等离子共振频移。对于块体金属材料,不同的金属表现出了不同的光泽,这意味着它们对可见光范围的各种颜色(波长)的反射和吸收能力不同。而其纳米材料的金属微粒几乎全部呈现出黑色,这是由于随着尺寸的减小,纳米颗粒对可见光的反射率降低而吸收率增强造成的。而在热力学性质上,1954年Takagi首先发现随着尺寸的减小,金属薄膜的熔点逐渐降低的现象。后来利用各种制备技术所得到的不同种类的金属纳米粒子、镶嵌粒子、金属薄膜及纳米线都发生了类似的熔点随尺寸的降低而逐渐减小的现象。此外,纳米材料的能隙、介电常数、导体向绝缘体的转变、磁的有序态向无序态转变、超导相向正常相转变,以及声子谱等都发现了与大块状态迥异的尺寸效应。

性能的变化必然是由结构的变化导致的,当纳米材料的尺寸减小到与光波波长,超导态的相干长度或电子的德布罗意波波长或透射深度等物理特征尺寸相当或更小的时候,晶体的周期性边界条件被破坏,纳米晶粒表面的原子密度减小而导致电子被限制在体积非常小的纳米空间里,电子的传输能力下降,平均自由程减小,从而电子的局域性和相干性得到增强。由于尺寸的减小使得纳米粒子空间内的原子数大大减少,宏观状态下的准连续能带消失而表现出分立能级,量子的尺寸效应就变得显著,这直接导致纳米尺寸体系下的光学、热学、磁学和电学性能表现出与常规材料不同的特性。此外,高的比表面积也使得纳米粒子表面的键态产生严重的失配状态,如增加的表面台阶和粗糙度,从而产生许多活性中心,这就明显地提高了固—固、固—液相转变和催化的效率,优化了化学反应路径,提高化学反应速度和定向等方面的效果。这就有助于我们更加深刻地理解和掌握内界面的结构与性能。此外,对于纳米尺寸的颗粒,其内部的位错消失,并且出现大量晶界,这又使纳米材料的强度和硬度显著提高,为合成和设计新型材料提供了可能的途径和方法。例如,在块体状态下在相图中完全不溶的两种或多种元素及化合物,在纳米尺度下则可以很容易地形成固溶体。利用纳米材料的这些特性,不但可以合成原子排序状态完全不同的两种或多种物质的复合材料,也可以把以前难以实现的有序—无序相、晶态相、金属玻璃、铁磁相与反铁磁相、铁电相与顺电相复合在一起组成具有特殊新性能的复合材料。

由于表面或者界面处的原子数比率增多,纳米材料的一个主要特征就是其大的表面体积比。纳米晶体的界面原子配位数减少,导致界面上出现大量断键,界面能增加,界面原子稳定性降低,从而表现出更高的活性。因此,与块体相比,纳米材料往往表现出更加优异的性能。这一特性虽然拓展了纳米材料在不同领域中的应用前景,但是这种高的表面体积比也提高了纳米粒子自身的能量,使它们经常处于亚稳状态,降低了热稳定性,这正为晶粒的生长创造了条件。对于纳米器件的应用,我们常常要求纳米材料能在更高的温度范围内保持其热稳定性和优良的性能。在较宽的温度范围内获得稳定优质的纳米结构材料是当前的科研工作者亟待解决的关键问题之一。目前的理论研究还主要集中在计算机模拟计算中,由于以原子的势函数为基础的模型模块普遍存在定量的问题,而且对于力场及经验函数的开发和选择等还不够完美,因此用热力学来描述纳米尺度的材料的特性就成为更加实际和方便的途径方法。研究纳米热力学,即纳米尺度材料中能量转换规律的问题,不仅可以拓展热力学在描述小尺度系统问题的理论基础和应用,而且对凝聚态物理研究纳米结构材料也是很好的支持和补充。这些都对当今纳米材料科学的研究具有非常重要的意义。

1.3 纳米材料的有序—无序转变

对于拥有 3d 电子结构的磁性金属元素与具有 5d 电子结构的金属元素组成的双金属化合物合金,比如 FePt, FePd, CoPt 等,由于具有较大的磁各向异性(MAE),经常被用来做超高密度磁记录设备。这主要是由其结构的变化而导致的,在温度的影响下,这些双金属化合物合金结构发生了有序—无序转变(order – disorder transition)。一般情况,在低温的时候,这些化合物形成了 CuAu –

类型($L1_0$)的有序结构,这种结构具有硬磁性;而当温度逐渐升高的时候,这种化合物合金形成了立方面心无序固溶体结构,这种结构只有较低的 MAE。描述 MAE 的指标是通过测量合金化合物的单轴磁性各向异性常数 K_u。受温度影响,由热浮动引起的磁化可以用热能 k_BT 来表示,能垒为 K_uV(其中,k_B 为波尔兹曼常量,T 为绝对温度,K_u 为磁化的各向异性常量,V 是磁域的体积)。在应用的时候,要求合金在磁化状态下所需的热稳定性满足 $(K_uV/k_BT) \geq 60$。但是在制备过程中,受限于操作的条件和环境,我们常常得到的是面心立方(FCC)的无序固溶体结构,这就要求我们通过退火技术来恢复双金属合金颗粒的有序结构。一份用第一原理计算 FePt 和 CoPt 合金结构的报告(Sakuma A,1994)指出,原子有序是合金结构拥有潜在较大 K_u 值的决定性因素,而合金结构的磁性对在制备过程中形成的有序状态非常敏感。随着现代科技的发展,制备尺寸更小的,特别是在尺寸小于 10 nm 的超密度磁存储器件成为当务之急。但是在制备尺寸更小的纳米颗粒时,尤其是 5 nm 以下的合金纳米颗粒,存在的一个技术难点,即由热浮动所引起的磁序方向反转而导致的超顺磁行为很难维持比较长的时间。因此研究有序—无序转变的临界点,研究尺寸对转变温度的影响就非常有意义。在实验上制备这种具有 $L1_0$ 结构的磁性纳米颗粒主要是通过化学合成和薄膜沉淀技术。S. H. Sun 等人(1989)提供了一套更简单的方法,即通过形核来制备具有更高分散性自组装磁性簇,他们的研究工作因此而受到了广泛的关注。此外,溅射和气相沉积法也都获得了具有 $L1_0$ 纳米结构的薄膜和纳米颗粒。

当前,研究有序—无序转变的理论工作还主要是通过计算机模拟计算。通过分子动力学和蒙特卡洛方法,从纳米尺度考虑几千甚至几百万个原子或分子的交互作用;通过有限元分析方法,从宏观角度考虑整个体系的变化趋势。而对于热力学模型,最近也有很多值得关注的工作相继展开。C. Q. Sun 和他的合作者(2006,2007)通过建立键序-键长-键强的相关性(BOLS)来描述颗粒表层失配原子的缺失的结合能,成功描述了合金结构的热稳定性,并用表面能和表面熵的概念揭示了有序—无序的转变过程,他们还结合模拟计算的结果与热力学模型描述了 CoPt 纳米线有序—无序相转变的机制。W. H. Qi 等人(2010)也发展出了他们的热力学模型(如图 1.1 所示),他们利用发展出的模型成功解释了由 Alloyeau 等人(2009)在实验上观察的 CoPt 纳米颗粒的有序—无序转变的尺寸及形状效应。根据他们的理论,有序—无序转变是从表面延伸到内核的相变过程,这意味着相变过程是由表面主导的。为了获得有序相,退火温度必须低于表面有序温度,这也为确定退火温度的上限确定了理论基础。

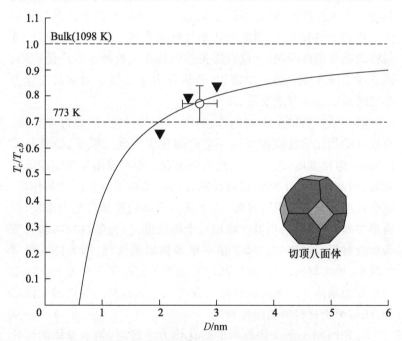

图 1.1　CoPt 切顶八面体(接近球形)纳米颗粒的有序—无序转变温度的尺寸函数

尽管以上提到的一些工作都很好地描述了有序—无序的转变过程,与计算机模拟的结果和实验结果都拟合得很好,但是还有很多关于有序—无序相变的细节问题并没有描述清楚,比如表面体积比对有序—无序转变的影响等,这就要求我们发展新的热力学模型来补充和深入我们对有序—无序转变过程及相关机制的理解。

1.4 纳米晶体材料熔化的尺寸效应

在凝聚态物质所表现出的所有现象中,物态的变化,尤其以物质的熔化和凝固最为重要,而19世纪以来兴起的热力学理论的根基正是建立在对熔化和凝固的深入研究基础之上的。由于纳米材料处于宏观尺度与微观尺度之间,也分别是量子力学计算与热力学分析的上限和下限,因此对纳米材料相变的研究,尤其是小尺寸材料的热稳定性问题,不仅直接关系到其在工程领域的广泛应用,而且对于深入理解热力学理论、分析热力学与量子力学及统计力学之间的联系也有重要意义。

熔化是材料最基本的现象,材料的很多性能,比如力学性能、物理性能和化学性能都可以与它的熔化温度建立联系,也就是工作温度/熔化温度(T_w/T_m)的比值的函数。由于该函数在室温下是常数,因此在室温下观察材料性能时往往忽略了这个函数的作用。但是在温度偏离室温的时候,这个函数的影响就显得尤为重要,特别是对高分子材料,当温度增加几十摄氏度之后,塑性材料就转变为橡胶材料。这也适合用于描述纳米材料的情况,因为随着尺寸的减小,纳米材料的熔点是降低的,相当于提高了工作温度。这也是在高温条件下纳米材料通常具有优于大块材料的性能的原因之一,比如纳米材料的高压延展性。

首先回顾一下大块材料的熔化热力学理论,纳米晶体的熔化热力学理论也是从大块材料的熔化理论基础之上而发展起来的。1910年,Lindemann在总结了爱因斯坦量子力学理论的基础上指出,只有在晶体中的原子振动均方根位移达到原子最近邻距离的某一临界比值时才能导致熔化的发生。之后,Bom(1939)提出了力学不稳定熔化判据。基本思想是在晶体发生熔化时,由于晶体具有抵抗剪切应力的弹性阻力,而液体不具有这一特性,从而切变模量消失。但是该理论的缺陷也很明显,那就是在接近熔化温度的时候,晶体的切变抗力并不为零。除此之外,其他的熔化理论也相继出现,比如位错熔化理论、缺陷熔化的判据、原子自由运动熔化准则等,但这些理论也都仅仅揭示了部分熔化现象。值得注意的是,理论研究发现,晶体的熔化是从表面熔化开始的,而熔化可以看作是液体在晶体表面生长扩散的过程。

Lindemann 准则虽然只是一个经验性的结论,但是自提出至今已经被科学界广泛接受,并成功应用于对晶体、非晶体和有机体熔化过程的解释和预测。对于纳米晶体材料,Lindemann 准则同样适用,随着尺寸的减小,表面体积比急剧增加,从而导致表面声子软化,表面原子均方位移移动更加自由,减小了熔化的势垒而降低熔化温度。继 1954 年 Takagi 发现了金属粒子熔化温度得到抑制之后,实验上又发现了高熔点的基体可以提高内嵌纳米晶体粒子的熔点的现象。大量的实验研究发现,纳米晶体粒子的熔化是与尺寸效应、维度效应和界面效应息息相关的。为了解释诸多实验现象,并预测熔化的发展趋势,学者们在理论上建立了很多热力学模型。其中,最早的理论表达式是由 Pawlow 在 1909 年提出的,并由 Hanszen 修正,即

$$T_m(D)/T_m(\infty) = 1 - 4V_s [\gamma_{sv} - \gamma_{lv}(\rho_s/\rho_l)^{2/3}]/[H_m(T)D] \quad (1.1)$$

式中,$T_m(D)$ 为纳米晶体的熔化温度;D 为尺寸,也就是有效直径;$T_m(\infty)$ 是大块的熔化温度;V_s 为大块晶体的摩尔体积;γ 为单位面积的表面能或界面能;ρ 是晶体的密度,下角标 s,l,v 分别为固态相、液态相及气态相的情况;$H_m(T)$ 为摩尔熔化焓的温度效应。对于立方金属晶体,可以近似地认为

$$\gamma_{sv} - \gamma_{lv} \approx \gamma_{sl} \quad (1.2)$$

假设 $\rho_s \approx \rho_l$,结合公式(1.1)和(1.2)有

$$T_m(D)/T_m(\infty) = 1 - 4V_s \gamma_{sl}/[DH_m(T)] \quad (1.3)$$

观察公式(1.3)我们发现,这与现今流行的经典 Gibbs-Thomson 熔化热力学公式(式 1.4)的结果是非常接近的。

$$T_m(D)/T_m(\infty) = 1 - 2(1/D_1 + 1/D_2)V_s\gamma_{sl}/H_m(T) \quad (1.4)$$

公式(1.4)中的参数 D_1 和 D_2 分别为椭球体晶体界面的两个主曲率半径,对于圆球状颗粒来说,$D_1 = D_2 = D$。

基于实验现象,当内嵌纳米粒子与高熔点基体的界面呈共格或半共格状态时,内嵌纳米粒子的熔点会随着尺寸的减小而提高。1977 年,Couchman 与 Jesser 建立了纯热力学模型来解释这种情况:

$$T_m(D)/T_m(\infty) = 1 - [6V_a(\gamma_{sM} - \gamma_{IM})/D - \Delta E]/H_m(T) \quad (1.5)$$

式中,$V_a = (V_s + V_1)/2$,V_1 是处于液态时的摩尔体积;γ_{sM} 与 γ_{IM} 分别是纳米粒子或相应的液体与基体的界面能;ΔE 表示纳米粒子的固体颗粒与同体积纳米小液滴之间的能态密度差。当忽略 ΔE 这一条件时,内嵌纳米粒子熔点的升高与降低直接取决于 $\gamma_{sM} - \gamma_{IM}$ 这一项。而这和颗粒与基体之间的界面情况直接相关,如果界面是共格或者半共格,$\gamma_{sM} - \gamma_{IM} < 0$,此时处于内嵌粒子表层与基体之间界面处的原子处于被束缚状态,内嵌粒子过热而导致熔点升高。反之,内嵌粒子表层处的原子没有受到束缚,与自由粒子相似,此时内嵌粒子的熔点降低。

过热的现象也曾用压力效应来解释,比如毛细管效应,因内嵌纳米粒子与基体之间热膨胀系数差异而产生的压力效应,以及由于体积变化而导致的压力效应等。但是这些模型只能预测不超过 6 K 的过热现象,这主要是由于只考虑了简单机械效应,而忽略了界面间的复杂的化学作用。

在解释纳米晶体熔化的现象时,公式(1.3)和(1.5)都在纳米晶尺寸大于 10 nm 的范围内取得了成功,不论是熔化抑制现象还是熔点升高的过热现象。但是对于更小尺寸的纳米晶体的熔化的预测,它们都失败了。这主要是由于公式(1.3)和(1.5)仅仅考虑了粒子表面的原子对熔点 $T_m(D)$ 的贡献,也就是随着表面体积比的增加,表面原子对整个纳米颗粒的性能的贡献。从数学角度看,这仅仅是一个一级近似;从物理角度看,忽略了表面原子的弛豫与重组,以及纳米晶表面熔化的因素。表面熔化是指在接近但未达

到大块熔化温度的时候,晶体表面的几个原子层便开始松动熔化的现象,如果考虑表面熔化,则 $T_m(D)$ 可以表示为

$$\frac{T_m(D)}{T_m(\infty)} = 1 - \frac{4V_s\left(\frac{\gamma_{sl}}{1-\delta/D} - \frac{\gamma_{lv}}{1-\rho_s/\rho_l}\right)}{DH_m(T)} \quad (1.6)$$

$$\frac{T_m(D)}{T_m(\infty)} = 1 - \frac{4V_s\gamma_{sl}[1-\exp(-\delta/\xi)]}{DH_m(T)(1-\delta/D)} -$$

$$\frac{V_s[(\gamma_{sv}-\gamma_{lv})-\gamma_{sl}(1-\delta/D)^2]\exp(-\delta/\xi)}{\xi H_m(T)(1-\delta/D)^2} \quad (1.7)$$

$$\frac{T_m(D)}{T_m(\infty)} = 1 - \frac{4V_s\gamma_{sl}DH_m(T)}{1-\delta/D} \quad (1.8)$$

式中,δ 和 ξ 分别表示表面处熔化液层的厚度及固液界面的相关长度。值得注意的是,当 $\delta \ll D, \delta \gg \xi$ 时,$\rho_s \approx \rho_l$,再通过公式(1.2),我们可以看出公式(1.6)—(1.8)的预测原理是与公式(1.3)一致的。也就是说,当纳米晶体的尺寸较大时,表面熔化的现象并不能改变纳米晶粒的整体熔化行为,但是当尺寸 $D < 10$ nm 或者其表面体积比 $hA/V > 10\%$(h, A, V 分别为原子的直径、表面积和体积)时,δ 的大小是与尺寸 D 相当的量,表面熔化变成一个不可忽视的因素,而公式(1.3)仅给出一个 $T_m(D)$ 与 $1/D$ 的线性关系,因此公式(1.6)—(1.8)的结果与公式(1.3)预测的差别很大。

在以纳米晶体的熔化是固体粒子内嵌在自身的液体中的假设为前提下,Semenchenko 提出了一个新的指数形式的热力学模型来解释熔化的尺寸效应,即

$$T_m(D)/T_m(\infty) = \exp[-4V_s\gamma_{sl}/H_m(T)D] \quad (1.9)$$

事实上,对于纳米材料熔化温度尺寸效应的描述,公式(1.9)在本质上是与公式(1.6)—(1.8)一致的,只是表达方式不一样而已。根据数学近似关系,当 x 较小时,$\exp(-x) \approx 1-x$ 成立,当尺寸不断增大时,公式(1.9)也与(1.3)相互一致的。所以,上面提到的这些热力学模型不仅可以在宏观尺度有效,也就是 $D \to \infty$ 时,$T_m(D) \to T_m(\infty)$,而且在当 $\infty > D > 10$ nm 时,也能做出很好的预测。注意到公式(1.6)—(1.8)中需要的这些参数要通过拟合实

验结果才能得到,而公式(1.9)则可以直接根据实验中的数据值得出结果,对于预测纳米晶体熔化的尺寸效应要方便得多。但是公式(1.9)既不能预测熔化的过热现象,也不能反映维度对熔化的影响。因为低维材料不同的维度具有不一样的表面体积比,从而影响了各自的熔化温度。实际上公式(1.9)所表达的只是一维情况,由于一维纳米线介于零维粒子和二维薄膜之间,所以在预测其他维度的材料时也能表现出一定的近似性。

我们注意到在上面所提到的模型中,固液界面能 γ_{sl} 项是一个很重要的热力学参数,但是实验上对这一参量很难直接测量。而纳米晶体的熔化过程又始于表面熔化(如图 1.2 所示),In 纳米晶体的表面能和界面能直接导致熔化的不同现象。

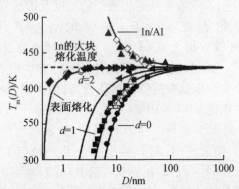

图 1.2 In 纳米晶体熔化温度 $T_m(D)$ 与 $T_m(D)$ 尺寸效应示意图(实线部分)

图 1.2 中包括纳米粒子、纳米线、纳米薄膜,以及 In 粒子内嵌在 Al 基体中、In 表面熔化等不同环境的情况。

所以通过理论计算来获取 r_{sl} 的值则变得非常有意义。根据 Gibbs-Thomson 公式,我们知道,

$$\gamma_{sl} = 2hS_{vib}(\infty)H_m(T)/(3V_sR) \quad (1.10)$$

式中,$S_{vib}(\infty)$ 是大块熔化熵 $S_m(\infty)$ 中的振动部分;R 为理想气体常量。如果忽略晶体的各向异性,事实证明,公式(1.10)对于定量计算离子晶体,金属和有机晶体的 γ_{sl} 是非常准确的。将公式(1.10)代入(1.3)中,则有

$$T_m(D)/T_m(\infty) = 1 - 8hS_{vib}(\infty)/(3RD) \quad (1.11)$$

虽然公式(1.11)似乎忽略了如表面粗糙化、表面熔化等使表面能降低的一些处理过程,但是由于 $S_{vib}(\infty)$ 表示的是在熔点处固体与液体振动熵的差值,其本身已经包含了表面弛豫的含义。然而,由于尺寸特别小而导致表面体积比大于20%时,公式(1.11)和(1.3)便失效了。比较有意思的是,虽然在上述公式中出现了 $H_m(T)$ 参数,但是在公式(1.11)中被消去了,也就是说,$H_m(T)$ 对 $T_m(D)$ 的影响意义并不大。

综上所述,虽然这些热力学模型对于纳米晶体材料的熔化现象都得到了很好的预测结果,但是当尺寸减小到 10 nm 以下时,$T_m(D)$ 会发生急剧的降低,但是还没有一种模型能够很好地解释这一尺度范围内的熔化现象。根据 Lindemann 熔化准则和 Mott 关于材料的熔化温度与振动熵关系,我们建立了关于 $T_m(D)$ 的不需要任何可调参数的新模型,该模型不但可以很好地解释整个尺寸范围内的金属、有机晶体及半导体熔化的尺寸效应,而且考虑到界面对熔化的影响,还完美地解释了对于镶嵌在基体内部的纳米粒子点熔化温度升高的过热现象。

对于纳米晶体熔化的尺寸效应,大量的热力学模型都给出了很好的解释和预测,但是我们对纳米尺度下热力学其他参量的变化研究还远远不够,而对这些参量尺寸效应的研究,有助于我们更好地对纳米尺度及低维材料中的热力学本质的理解。此外,对于纳米尺度和低维材料的其他性能的尺寸效应的研究依旧很匮乏。因此,我们所建立的这个模型是否也适合描述这些领域,以及如何拓展到这些领域也是亟待解决的问题。

1.5 固态相变的尺寸效应

拓展纳米材料的应用领域不仅依赖于其制备技术的发展和完善,而且对其结构、性能的理解和认识的程度也至关重要。材料在制备及应用的过程中,在不同的外界条件,例如温度和压力的影响下经常会发生固态相变,即从某一固相转变成另外一固相。在实际制备和应用纳米材料的过程当中经常会出现亚稳态和非亚稳态的微观组织结构,但是对纳米尺度的相变,当前的理论研究和相图还很少,因此对纳米材料相变的研究就显得非常有意义,有必要在理论上就固态相变的尺寸效应进行进一步的拓展和探讨。此外,由于材料的力学性能直接影响材料器件的稳定性,而纳米尺度材料的晶格、弹性性能等都与其在大块尺寸时有较大差异,因此对纳米材料的应力—应变研究也是一项意义深远的工作。

材料热力学是专门研究材料的稳定性和能量变化的一门科学,在一定条件下研究材料的相变是材料热力学的本质问题。由于材料的结构决定了材料的性能,而材料的结构又受到其本身的成分,外部的压力和温度等因素的影响。只有弄清了外部条件、材料结构、材料性能之间的关系,我们才能更方便地制作出想要的材料的结构和更好的应用材料。所以研究材料的结构及其相变在科学领域就显得非常重要,研究材料在相平衡下的相变是描述材料性质的第一步。对于纳米尺度的材料,由于其在自然界中都是处于亚稳状态的,尤其在高温高压条件下很难判断纳米材料是哪种相结构,因此,用热力学的方法来建立理论模型描述纳米相变的相图就要方便实用得多。

最早开始研究相变的文章是始于19世纪关于固态、气态和液态物质三态变化的工作。经过几十年的研究发展,到现在研究的热点问题开始转向更加关注材料的铁电转变、磁性转变、超导转变,以及其他结构的相变,例如前面提到的有序—无序转变、马氏体转变、玻璃相变等。

描述材料的内部结构的因素很多,它们之间的关系也很复杂,如能带结构、电子结构、组成成分、结合态、对称结构等,而这些因素又都受到外部条件比如压力/应变、温度、电场、磁场等的深刻影响。这些外部因素中的每一个发生变化,都可能影响材料的内部结构参数的变化。热力学上对大块尺寸相变的研究,其条件主要设定在压力、温度和成分对相变的影响,因为这几个条件也是我们生活当中最常遇到的情况,这在经典热力学中利用吉布斯自由能公式都得到了很好的描述。而对于纳米尺度材料,新增加的自由度－尺寸(D)大大扩展了材料热力学的研究范围,成为研究材料热力学相变中的另外一个重要参量。由于 D 的变化涉及材料的表面体积比的变化,其中又掺杂着材料维度(纳米粒子、纳米线、纳米管及纳米薄膜)和形状(球体、立方体及多面体)的变化。特别是当尺寸 D 减小到 1～100 nm 时,这些因素的共同作用使得纳米材料在结构和热力学稳定性方面表现出与大块时完全不同的状态。在不同的温度和压力条件下纳米材料的结构发生着不同的变化,使得其吉布斯自由能 G 达到最小的稳定状态。研究纳米材料热力学相变的尺寸效应在现代工程技术中的地位将变得越来越重要。

目前,有三种常用的基本的热力学方法来研究纳米材料领域的尺寸效应,它们分别是以温度的起伏、Tsallis 熵和描述小尺寸体系的 Laplace-Young 方程为基础的。第一种方法仅从热力学第一定律出发,而没有考虑其他的热力学关系,是一种普适的用来处理纳米尺度的热力学模型;第二种方法是基于 Tsallis 的广义玻尔兹曼－吉布斯统计热力学而发展的通过热力学附加属性(特别是熵的贡献)来表现,其中也包括了非广延的纳米热力学体系;第三种方法是考虑了内压力 P_{in} 贡献,这种方法也经常用来描述大块尺寸的情况,因为虽然内压力产生的原因不同,但是它们对材料的结构的影响本质都是一样的。这些方法是从不同的角度来描述微观世界的现象,它们对研究微观领域热力学的影响都是深远的。但是由于纳米科学及纳米技术的迅猛发展,发展一个简单、定量的没有任何可调整自由参数的普适热力学模型来描述纳米材料的相变就显得非常重要。

1.5.1 纳米固态相变热力学的研究现状

固态相变研究始于 1965 年 Garvie 对 ZrO_2 马氏体相变的研究,他确定了在室温条件下尺寸效应对纯 t-ZrO_2 的稳定性所起到的作用,并指出正是大块稳定相与纳米粒子的亚稳相之间的表面能差 $\Delta\gamma$ 使得大块时的亚稳结构在尺寸减小到纳米级时能够稳定存在。

通过测量和对比 m-ZrO$_2$ 相和 t-ZrO$_2$ 相的表面能，得出在室温条件下发生从 m-ZrO$_2$ 相到 t-ZrO$_2$ 相转变的临界尺寸为 D_c = 30 nm，也就是说，在室温 298 K 时，对于尺寸小于 D_c 时的纳米粒子，其稳定结构为 t-ZrO$_2$ 相。

后来，Banfield 等人（2005）通过热力学和动力学方法研究了金红石结构 r-TiO$_2$ 与锐钛矿 a-TiO$_2$，以及六方纤维锌矿 w-ZnS 与立方闪锌矿 s-ZnS 之间的相变，发现表面能 γ 和表面应力 f 对总能量 G 的增加都有贡献。由于当时还无法确定表面应力 f 的数值，因此他们假设 $f \approx \gamma$。

直到 Barnard 和他的合作者们（2004）发展出了一套可以用来计算任何形状纳米材料热力学稳定性的热力学模型。该模型考虑了纳米晶体的各个表面、各个顶点及各个棱边对总能量的贡献。他们也对表面应力 f 的作用做了深入的探讨。他们用该模型描述了各种形状的 a-TiO$_2$ 与 r-TiO$_2$ 纳米粒子的最小能量，以及表面经过处理过的 a-TiO$_2$ 和 r-TiO$_2$ 发生相变的临界尺寸。同时，他们也计算了 t-ZrO$_2$ 与 m-ZrO$_2$ 纳米晶体最小能量和稳定性，以及 ZnO 纳米带的生长和发生的相变。实验结果也证实了他们模型预测的准确性。

此外，Yang 等人（2004）通过引入 Laplace-Young 方程，建立了相应的热力学模型描述固体相变的尺寸效应，成功地预测了金刚石的形核，以及当环境条件发生变化时其结构随之发生的一系列变化，该模型能够很好地解释同一种结构能够沿着不同的生长方向发生一系列的转变。

近几十年的研究报告中都揭示出纳米粒子的结构和稳定性都异于大块时的状态。在本书中提到的稳定相是在不考虑尺寸参数的影响下在常温常压条件中可以稳定存在的相，而亚稳相则指在常温常压条件下不稳定，当尺寸减小超过某一临界尺寸时稳定存在的相。在我们的工作中，一个没有任何可调参数的简单热力学模型建立起来系统地描述纳米尺度的固态相变，同时也考虑了表面能及表面应力的贡献。

1.5.2 热力学模型的建立基础

吉布斯自由能是衡量相变过程中两相竞争稳定性的一个标准,吉布斯自由能最小的相将稳定存在,而吉布斯自由能比较大的相在一定条件下由于动力学因素可以存在的相则为亚稳相。

在平衡条件下,材料的结构由 i 相转化成 j 相时,它们各自的吉布斯自由能相等,也就是

$$G_i = G_j \tag{1.12}$$

表面能 γ 所诱导的吉布斯自由能,也就是表面自由能又可以表示为

$$G_s = \gamma A \tag{1.13}$$

则 i 相和 j 相的吉布斯自由能差可表示为

$$\Delta G = G_v^i - G_v^j + A^i \gamma^i - A^j \gamma^j \tag{1.14}$$

式中,G_v 表示标准体积吉布斯自由能。表面体积比 q 的公式又可以写成

$$q = A/V = 2(3-d)/D \tag{1.15}$$

代入公式(1.14)中则有

$$\Delta G = \Delta G_v + 2(3-d)(V^i \gamma^i - V^j \gamma^j)/D \tag{1.16}$$

式中,$\Delta G_V = G_v^i - G_v^j$。当 $\Delta G = 0$ 时,i 相的纳米晶体与 j 相的纳米晶体在给定的温度压力条件下能量相等,系统达到平衡状态。

1.5.3 纳米晶体表面应力对固态相变的影响

上述的讨论结果中并没有涉及表面应力的影响,但是在实际情况中,由于纳米晶体表面原子失配导致表面键收缩,从而产生内压力 P_{in},这个内压力是由表面应力 f 引起的,当晶体的尺寸减小到纳米尺度时,P_{in} 对纳米晶体的稳定性的影响是不能够忽略的。P_{in} 可以表示为

$$P_{in} = 4f/D \tag{1.17}$$

对于某一指定尺寸的纳米晶体,当受到外力引起弹性变形的时候,其体应力和表面应力同时做功,在尺寸比较大的时候,其体应力的影响要远远大于表面应力。但是随着尺寸的减小,表面体积比急剧增加,从而使得表面应力的贡献逐渐增大,这时表面应力

的影响就不能不考虑了。定义在新产生的表面或界面面积 dA 上做的功为 dW,也就是

$$dW = \gamma dA \qquad (1.18)$$

注意:γ 在这里的含义为单位面积上的过剩自由能。而对于固体表面,在弹性应变张量 ε_{ij} 与表面应力张量 f_{ij} 影响下所做的功,即其表面总能量的变化可以表示为

$$d(\gamma A) = A f_{ij} d\varepsilon_{ij} \qquad (1.19)$$

由于 $d(\gamma A) = A d\gamma + \gamma dA$,又 $dA = A\delta_{ij}d\varepsilon_{ij}$(其中 δ_{ij} 为 Kronecker 参数),因此表面应力张量 f_{ij} 就可以定义为

$$f_{ij} = \gamma \delta_{ij} + \partial \gamma / \partial \varepsilon_{ij} \qquad (1.20)$$

对于一般的表面,由于其对角线上的张量部分为零,其主轴部分代表着表面应力张量;而对于具有三轴或更多对称轴的表面的情况,由于对角线部分相同,则表面应力可以定义为

$$f = \gamma + \partial \gamma / \partial \varepsilon \qquad (1.21)$$

其中,ε 表示的是表面处的晶格应变。

考虑表面应力对系统总能量的贡献,此时公式(1.16)则可以表示为

$$\Delta G = \Delta G_v + 6(V^i \gamma^i - V^j \gamma^j)/D + 4(f^i V^i - f^j V^j)/D \qquad (1.22)$$

从公式(1.22)中可以看出,ΔG_v 仅仅是由温度的变化而引起的能量差,与尺寸是无关的,而后两项分别代表了表面能 γ 与表面应力 f 对整个系统能量变化的贡献。

第 2 章 形状、尺寸和维度依赖的相变温度

2.1 磁性二元合金的有序—无序转变温度

2.1.1 磁性二元合金有序—无序转变背景简介

近些年来,磁性纳米颗粒获得了广泛的关注和科学研究,尤其是当尺寸小于 10 nm 时在超磁存储设备上的潜在应用。因为磁性颗粒主要是应用于高密度存储媒介,所以为了能够克服在特定磁性方向上的热浮动,这种颗粒必须具备更高的磁各向异性。这种性质可以用热稳定因子来评估,也就是保证磁化热稳定性和热能 ($K_u V/k_B T$) >60。其中 K_u, V, k_B, T 分别表示磁晶的各向异性常量、磁域的体积、波兹曼常量和绝对温度。为了发展能够满足这个关系式的新材料,一些拥有比较高的 K_u 值和大的矫顽力的二元合金,比如 FePt,CoPt,FePd,以及一些拥有 CuAu I ($L1_0$) 结构类型的化学有序结构合金在计算机动力学模拟上或者试验中都受到了很大的关注。然而,在制备这些合金的时候,通过一些常用的合成方法,比如湿化学法或者物理蒸发沉淀法,经常出现一些无序的面心立方 (fcc) 固溶体结构,而这种结构拥有较低的磁各向异性。通过研究发现,这种二元合金通常只有在低温下才能产生 $L1_0$ 结构,当温度逐渐增加并超过临界温度时,出现 fcc 面心立方无序结构。这个从低温 $L1_0$ 有序结构向高温 fcc 无序结构转变的临界温度,就被定义为有序—无序转变温度,一般用 T_{OD} 表示。为了恢复原子有序和优良的磁学性质,需要提高合金的最高退火温度(T > 873 K)以保证 $L1_0$ 有序结构的稳定性。然而,对于尺寸小于 5 nm

的合金颗粒,这个过程在实验上特别难实现。为了能够在理论上更好地理解尺寸对有序—无序转变的影响,很多针对 FePt,Cu_3Au,FePd 和 CoPt 二元合金的研究工作已经陆续展开。研究发现,T_{OD} 直接与颗粒的表面体积比 δ 相关联,并且发展出能够有效测量小尺寸颗粒(一般直径小于 4 nm)有序参数的方法。表面的低配位状态引起的表面无序趋势更有利于相转变的发生。例如,对 CoPt 纳米合金,有相关报道(Alloyeau,2009)指出,当颗粒尺寸在 2~3 nm 时,T_{OD} = 773~923 K,比相应的大块值(1098 K)小很多。低配位对磁性各向异性的影响最近也在 Co 纳米颗粒中发现。2006 年,J. S. Kim 在高温退火处理 CoPt 纳米颗粒时,发现 CoPt 颗粒的成分也展现出尺寸依赖效应,这主要是由 Co 原子具备较高的蒸发速率而导致的。需要注意的是,由于非平衡动力学因素的影响,当 D = 2.4 nm,且温度低于 T_{OD} 时,原子无序结构一样有可能存在。另外,当纳米颗粒在尺寸 D = 3.5 nm,温度在 873 K 或者以下时,很多实验观察不到有序相的存在,这也都是由动力学因素导致的。最近,纳米颗粒的形状对 T_{OD} 的影响也引起了科学家的兴趣,很多实验和计算机模拟的工作已经相继展开。结果显示,纳米颗粒的表面体积比对调整 T_{OD} 起到了非常重要的作用,即使仅在一维方向上减小,也能对 T_{OD} 起很大的抑制作用。并且,通过小角和广角散射方法探测到在 1.5~3.5 nm 范围内不同的退火温度也能导致 CoPt 纳米颗粒不同结构的产生。此外,相关的理论方法也得到了发展。

在本节中,我们根据以前提出的熔化热力学模型,通过引入相应的附加参数来描述形状、尺寸和维度对 T_{OD} 的影响,发现形状对 T_{OD} 的影响要比维度弱,相关的实验结果也证实了我们的预测。

2.1.2 二元合金有序—无序转变温度的尺寸与形状函数模型

根据 Lu(2008)的模型,T_{OD} 可以直接关联德拜温度 θ 和熔化温度 T_m,即

$$\frac{T_{OD}(D)}{T_{OD}(\infty)} \approx \frac{\theta(D)}{\theta(\infty)} = \left[\frac{T_m(D)}{T_m(\infty)}\right]^{1/2} = \exp\left(-\frac{S_{vib}}{3R}\frac{1}{D/D_0 - 1}\right)$$

(2.1)

式中, D 表示纳米颗粒或纳米线的直径,也可表示薄膜的厚度; ∞ 代表大块尺寸; R 是理想气体常量; D_0 是当所有原子都处在表面时的临界直径,可以用维数 d 和最小原子距离 h 表示为 $D_0 = 2(3-d)h$; $d=0,1,2$ 分别对应球状颗粒、线和薄膜,也就是 $D_0 = 6h, 4h, 2h$; S_{vib} 是大块熔化熵 S_m 中的振动部分,也就是振动熵。因为在有序—无序转变过程中二元金属合金的键的化学性质并未发生变化,所以电子熵的贡献可以忽略不计,即 $S_{vib} \approx S_m - S_{pos}$, S_{pos} 是位置熵的贡献。

为了讨论形状的影响,我们引入形状因子 λ,定义为其他形状的表面体积比 δ 对基本形状(相对颗粒为球形、相对线为圆柱形)的比值。将附加的参数引入公式(2.1),我们得到

$$\frac{T_{OD}(D)}{T_{OD}(\infty)} = \exp\left(-\frac{\lambda S_{vib}}{3R}\frac{1}{D/D_0 - 1}\right) \qquad (2.2)$$

假设不同形状的颗粒都具有相同的堆积密度,则 λ 可以表示为

$$\lambda = \delta_S/\delta_B = (A_S/A_B)(V_B/V_S) \qquad (2.3)$$

式中, A 和 V 分别表示表面积和体积,下角标 S 和 B 代表其他形状和基本形状。

表 2.1 是计算过程所需参数表。

表 2.1 计算过程所需参数

颗粒	h/nm	$H_m(\infty)$/(kJ·mol^{-1})	$T_m(\infty)$/K	S_m/(J·mol^{-1}·K^{-1})	S_{vib}/(J·mol^{-1}·K^{-1})
Fe	0.248	13.8	1811	7.62	
Pt	0.278	19.7	2041	9.58	
Co	0.25	15.2	1768	8.59	
FePt	0.277			8.6	2.84
CoPt	0.264			9.085	3.32

注:① FePt 和 CoPt 的 h 值定义为 Fe 或者 Co 和 Pt 的最近邻原子间的距离,其值来源于参考文献[155]和[160],其他参数则来源于文献[161]。

② 对于元素, S_m 可以通过 $S_m = H_m/T_m$ 直接得到;而对于 FePt 和 CoPt 合金,由于 H_m 和 T_m 的值无法直接得到,可以近似处理为 S_m(FePt 或 CoPt) = $[S_m(\text{Fe 或 Co}) + S_m(\text{Pt})]/2$。

③ 位置熵的贡献可以通过公式 $S_{pos} = -R[x\ln x + (1-x)\ln(1-x)]$ 近似处理,其中 $x=0.5$ 表示 Fe 和 Co 的原子比。

2.1.3 二元合金有序—无序转变模型和实验结果的对比与分析

图 2.1 两幅子图分别描述了由公式(2.2)和(2.3)确立的 FePt 和 CoPt 不同形状颗粒(球体(Sph)、切顶八面体(To)和十面体(Dh))的 $T_{OD}(D)$ 函数,及其 MD 分子动力学模拟(空心符号)和实验的结果(实心符号)的对比,用到的参数已在表(2.1)中给出。

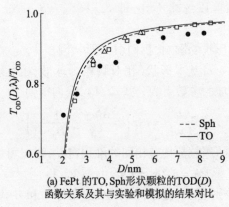

(a) FePt 的 TO,Sph 形状颗粒的 TOD(D)
函数关系及其与实验和模拟的结果对比

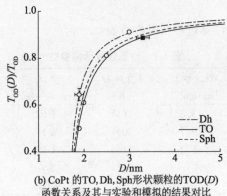

(b) CoPt 的 TO,Dh,Sph 形状颗粒的 TOD(D)
函数关系及其与实验和模拟的结果对比

图 2.1 模型计算结果

图中,实心点●,■表示实验对于球形的结果,空心点是模拟计算得到的切顶八面体(□,△,○)和十面体(◇)形状的结果。

从图 2.1 中可以很清楚地看到,$T_{OD}(D)$ 是随着尺寸 D 的减小而降低的,特别在尺寸小于 5 nm 的范围内,变化尤为明显。根据公式(2.3),不同形状的颗粒拥有不同的形状因子,也就是,对于十面体、球形、切顶八面体,形状因子分别为 $\lambda = 0.77$,$\lambda = 1$ 和 $\lambda = 1.12$,$T_{OD}(D)$ 函数有着明显的差别。从图 2.1b 中可以看到,形状为切顶八面体的 CoPt 颗粒的有序—无序转变温度 $T_{OD}(D)$ 明显小于十面体结构,相应的模拟计算结果也证实了这一点。但是对球状的 CoPt 和 FePt 的 $T_{OD}(D)$ 值明显大于实验结果,这很可能是由实验条件所致。比如非晶 Al_2O_3 和 MgF_2 基体的存在大大限制了界面处原子的运动,因而抑制了界面处混乱的趋势,从而降低了有序—无序转变温度。然而我们的模型预测与计算机模拟的结果符合得很好,这主要是由于模拟的环境更趋于理想化的情况,与我们模型适用的条件比较接近。对比实验结果、模拟的数据及模型的预测,发现各种形状的转变温度遵循如下顺序:$[T_{OD}(D)/T_{OD}]^{Dh} >$ $[T_{OD}(D)/T_{OD}]^{Sph} > [T_{OD}(D)/T_{OD}]^{TO}$。这种现象同样可以通过简化热力学模型来理解。根据数学近似,当 x 足够小时(一般是 $x < 0.1$ 或 $D > 20h$),$\exp(-x) \approx 1 - x$。据此,公式(2.2)可以进一步简化为 $(T_{OD}(D,\lambda)/T_{OD}(\infty)) \approx 1 - (2\lambda S_{vib} h/RD)$。从这个简化式中可以看到,$T_{OD}(D,\lambda)$ 是随着 λ 的增加和 D 的减小而减小的。因为 D 的变化范围是从几纳米到几百纳米,而 λ 主要的变化范围在区间(0,2)内,所以 D 是对 $T_{OD}(D,\lambda)$ 的变化起主要作用的。从图 2.1b 中还可以看到,当 $D = 3$ nm 时,切顶八面体的模拟数据明显偏离于我们预测的结果,这主要是由于当尺寸减小到一定范围时,尤其当 $D < 5$ nm 时,由于表面存在大量的断键,每增加或者减少一个原子都会导致结构的明显变化,不同结构之间的竞争使得模拟结果偏离。例如,D. Liu(2009)称 Ag 团簇包含 38 个原子时,切顶八面体的结构是最稳定的;而当原子数达到 75 个和 101 个时,十面体的结构变得更加稳定。对于二元 CoPt 合金来说,这种情况也难以避免。原子结构越稳定,原子从表面迁移出去所需要克服的势垒就越大,也就需要更高的温度,因此 $D = 3$ nm 时切顶八面体的结构拥有更高的温度就很好理解了。

图 2.2 为根据公式(2.2)和文献[39],对不同维数 d 的 T_{OD} 进行的比较。

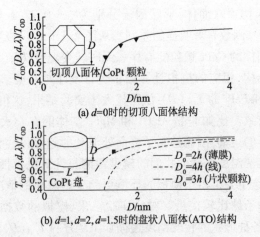

(a) $d=0$ 时的切顶八面体结构

(b) $d=1, d=2, d=1.5$ 时的盘状八面体(ATO)结构

图 2.2 对不同维数的 T_{OD} 的比较

现在考虑附加的形状效应对非球状纳米颗粒有序—无序转变的影响。在文献[39]中,CoPt 纳米颗粒有两个微观形貌,一个是尺寸为 3 nm 的各向同性的切顶八面体结构,另一个是各向异性的 1.5 nm 厚、4 nm 宽的平面压扁的盘状八面体结构(ATO),如图 2.2 所示。对于前者,其形状因子经过计算为 $\lambda = 1.12$,此时对于球状颗粒 $D_0 = 6h$,我们看到模型与实验结果符合得很好。而对于 ATO 结构,此时 $D/L = 1.5/4 = 0.375, 2h < D_0 < 6h$,是一种薄片状结构。这种结构的 T_{OD} 数值与球状颗粒 $D = 3$ nm 时是一样的。此时,$A = 2\pi(L/2)^2 + \pi LD$,并且有 $V = \pi(L/2)^2 D$。在第一个方程中当 $L \gg D$ 时假设 $D \approx 0$,结果 $\delta = A/V \approx 2/D$。将这些关系式代入公式(2.3)中,此时形状因子为

$$\lambda = (L+2D)/L = 1 + 2(D/L) \qquad (2.4)$$

将 $D = 1.5$ nm 和 $L = 4$ nm 代入公式(2.4)中,得到 $\lambda = 1.75$。对于 ATO 结构,因其同时具有薄膜和线的结构特征,可以近似假设 $D_0(薄片) = [D_0(2h 为薄膜) + D_0(4h 为线)]/2 = 3h$,这与实验结果精确拟合,从图 2.2 b 可以直接观察到。所以对于薄膜、片状颗

粒和线,有趋势 T_{OD}(薄膜) > T_{OD}(片状颗粒) > T_{OD}(线)。由公式(2.2)和公式(2.4)知,对于这种纳米薄片的 T_{OD} 随着尺寸 D 的减小和比率 D/L 的增加而下降。D 对于有序—无序转变的抑制作用要比 D/L 的作用强得多。

2.1.4 本章小结

根据熔化温度的尺寸与维度效应模型,我们发展了可以描述形状与维度效应对有序—无序转变温度 T_{OD} 的影响的模型。该模型不仅能够描述标准形状的纳米晶体的 T_{OD},而且能够准确描述压扁的 ATO 结构的 T_{OD},不同维度和形状的 T_{OD} 遵循 T_{OD}(薄膜) > T_{OD}(片状颗粒) > T_{OD}(线)的顺序。结果显示,ATO 结构的 T_{OD} 随着 λ 的增加和 D 的减小而下降,而 D 的作用要比 λ 的影响强得多。

2.2 Au-Pt 双金属纳米颗粒的熔化温度

2.2.1 Au-Pt 二元金属纳米晶体的熔化背景简介

与大块材料相比,由于金属纳米颗粒具有大的表面体积比,经常表现出更加良好的物理、化学和电学性质。尤其二元金属纳米颗粒,往往展现出比单金属颗粒更加优异的性能。对二元金属纳米颗粒,其物理、化学性质主要取决于颗粒的尺寸、成分、原子分布及形状,结果导致这些二元金属颗粒在异质催化、磁性设备、传感器、光电等领域有很大的应用前景。在各种物理性质中,熔化温度是一个非常重要的参量,尤其影响颗粒的形核质量,以及对材料进行精确的设计、控制和应用。因此,更好地理解尺寸、成分、原子分布和形状对熔化温度 T_m 的影响对于材料的合成和实际应用都有着非常重要的意义。

在各种二元合金纳米颗粒中,Au-Pt 纳米颗粒因其在燃料电池如乙醇的氧化反应和氧气的还原反应中具有独特的电催化特性,吸引了科学家们极大的研究兴趣。最近,具有不同尺寸、成分、元素分布和微观形貌的二元 Au-Pt 纳米颗粒已经在实验中相继合成(R. W. Shi,2011;W. H. Qi,2010;F. Y. Chen,2007)。然而,研究

T_m 用到最多的方法主要是分子动力学模拟。这主要是由于分子动力学不仅能够提供精确的能观察到颗粒变化的微观细节,而且可以任意设计颗粒的尺寸、形状、成分和结构。H. B. Liu(2008)的分子动力学模拟显示,Au-Pt 合金的 T_m 随着颗粒的尺寸减小而降低,这与单金属颗粒的变化趋势是一致的。此外,T_m 同样依赖于设计的结构、成分和形状。在其他合金颗粒的熔化过程中同样可以观察到类似的结果。尽管也有很多理论模型考虑了纳米颗粒的熔化的尺寸效应,但是没有相关的模型同时考虑尺寸、成分和形状的依赖效应。为了更深入地理解熔化现象,以及指导实践应用,用热力学的方法来发展一套能够同时描述尺寸、成分和形状效应对熔化温度 T_m 的理论模型将很有意义。

在本书中,依据结合能熔化尺寸效应的模型,我们发展了一个新的数学解析式来描述尺寸、成分和形状对 Au-Pt 纳米颗粒熔化温度 T_m 的影响,其预测结果与最近的分子动力学计算符合得很好。

2.2.2 二元合金熔化温度的尺寸与形状函数模型的建立

根据尺寸、形状和维度影响的结合能模型,相应的熔化模型为

$$\frac{T_m(D,\lambda)}{T_m} = \left(1 - \frac{1}{12D/D_0 - 1}\right)\exp\left(-\frac{2\lambda S_b}{3R}\frac{1}{12D/D_0 - 1}\right) \quad (2.5)$$

式中,S_b 是大块固—气转变熵,可以根据大块蒸发焓和大块蒸发温度来计算求得,即 $S_b = \Delta H_b/T_b$。

对于二元合金颗粒,应该考虑成分效应。观察公式(2.5)发现,只有 D_0 和 S_b 是受成分影响的参数,也就是公式(2.5)又可以写成

$$\frac{T_m(x,D,\lambda)}{T_m} = \left[1 - \frac{1}{12D/D_0(x) - 1}\right]\exp\left[-\frac{2\lambda S_b(x)}{3R}\frac{1}{12D/D_0(x) - 1}\right]$$
$$(2.6)$$

式中,D_0 可以写成由成分决定的原子最近距离的形式:

$$D_0(x) = 2(3-d)h(x) \quad (2.7)$$

为了获得这些成分决定的参数,这里引入经验公式(Fox 公式)来近似处理。由公式(2.7)可以发现,$D_0(x)$ 直接与最近原子距离

$h(x)$相关联,根据Fox方程,有

$$\frac{1}{h(x)} = \frac{1-x}{h(0)} + \frac{x}{h(1)} \quad (2.8)$$

式中,$h(0)$ 和 $h(1)$ 分别表示大块金属晶体在成分 $x=0$ 和 $x=1$ 时其最近原子间的距离。这个假设是合理的,因为最近的研究发现,对于大块 Au-Pt 合金,其晶格参数表现出类似的成分决定的关系。结合公式(2.7)和公式(2.8),$D_0(x)$ 得以确定。由于 $S_b(x)$ 在实验上很难获得,因此同样用 Fox 方程来近似处理。

$$\frac{1}{S_b(x)} = \frac{1-x}{S_b(0)} + \frac{x}{S_b(1)} \quad (2.9)$$

类似的处理方法在其他文献中应用并且得到了很好的结果。

结合公式(2.6)、公式(2.8)和公式(2.9),当相应大块参数确定后,一个简单的数学解析式就可以建立起来并描述尺寸、形状和成分依赖的熔化温度。

表 2.2 中列出了计算过程中所需主要参数,其中 h、T_m、T_b 和 ΔH_b 的值都引自文献[161]。

表 2.2 计算过程中需要的主要参数

金属	h/nm	T_m/K	T_b/K	$S_b/(\mathrm{J \cdot mol^{-1} \cdot K^{-1}})$	$\Delta H_b/(\mathrm{kJ \cdot mol^{-1}})$
Pt	0.2775	2041.4	4098	119.57	490
Au	0.2884	1337.3	3129	105.46	330

注:S_b 的值通过公式 $S_b = \Delta H_b / T_b$ 计算。

2.2.3 Au-Pt 二元合金熔化温度理论模型和实验结果的对比与分析

图 2.3 为文献[179]中采用的二十面体结构的 Au-Pt 纳米颗粒 $T_m(x,D,\lambda)$ 函数和 MD 分子动力学计算得到的结果的对比,其中形状因子根据公式(2.3)计算得 $\lambda = 1.40$。

图 2.3 描述的是所建模型的预测和利用相应分子动力学计算 Au-Pt 纳米颗粒模型结果的对比,需要的相应参数在表 2.2 中列出。图中引用的数据是基于 2.5 nm 左右的二十面体核壳结构小团簇,这种结构在这个尺度范围内被认为是最稳定的结构。从图

中可以看出,随着 Pt 成分 x 的增加,熔化温度 T_m 是逐渐升高的。这主要是由于在低熔点元素 Au 中(当 $x=0$ 时 Au 的熔点为 1337.3 K)添加高熔点元素 Pt 所导致的(当 $x=1$ 时 Pt 的熔点为 2041.4 K)。此外,Pt 的添加而导致的 T_m 提高也可以通过分析最近原子间的相互作用能来理解。我们知道,熔化温度和结合能是成正比关系的(结合能可以用蒸发焓 ΔH_b 来评估)。观察表 2.2 可以发现,蒸发焓有如下关系:$\Delta H_b(Pt) > \Delta H_b(Au)$。这意味着两个最近邻 Pt-Pt 原子之间的相互作用能要大于两个最近邻 Au-Au 原子之间的相互作用,而且最近邻 Au-Pt 原子对的相互作用也要强于最近邻 Au-Au 之间的相互作用。所以随着 Pt 成分的加入,合金结构的稳定性要比纯 Au 纳米颗粒的稳定性大。另外,这种稳定性的趋势同样可以通过对比大块固溶体成分依赖的形成热来理解。如文献[197]中,当 Pt 的成分 x 从 25% 增加到 75% 时,纳米颗粒的形成热从 0.048 eV/原子增加到 0.065 eV/原子。这就意味着 Pt 元素的添加使得合金团簇的结构更加稳定,从而提高了其熔化温度。图 2.3 显示,所建模型可以在整个成分范围内与相应的分子动力学模拟结果完美拟合。这也验证了 Fox 方程可以很好地处理 Au-Pt 固溶体合金的成分效应。图中的熔点误差主要是来源于文献[179]中计算所使用的不同方法。

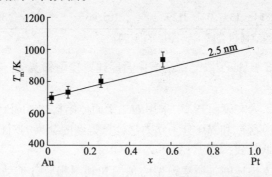

图 2.3 二十面体结构的 Au-Pt 纳米颗粒模型计算结果对比

图 2.4 是根据参考文献[181]所画的立方八面体(Cubo)、二十面体(Ih)、十面体(Dh)等不同形状的结构示意图。结合形状因子

计算的结果在图 2.5 中得到很好的体现。

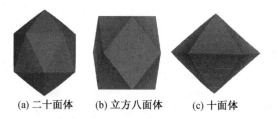

(a) 二十面体　　(b) 立方八面体　　(c) 十面体

图 2.4　根据文献[181]所画的二十面体、立方八面体和十面体

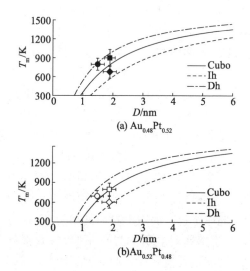

图 2.5　不同成分和形状的模拟计算结果和利用公式(2.6)计算的 Cubo、Ih 和 Dh 结构的 $T_m(x,D,\lambda)$ 函数结果对比

图中，实心符号■、◆和●分别表示 $Au_{0.48}Pt_{0.52}$ 的 Cubo、Ih 和 Dh 结构的分子动力学模拟结果；空心符号□、◇和○分别表示 $Au_{0.52}Pt_{0.48}$ 的 Cubo、Ih 和 Dh 结构的分子动力学模拟结果。

Au-Pt 合金的成分分别为 $Au_{0.48}Pt_{0.52}$ 和 $Au_{0.52}Pt_{0.48}$。当 Pt 的成分 x 增加时，相同直径下 $Au_{0.48}Pt_{0.52}$ 的熔点 T_m 要比 $Au_{0.52}Pt_{0.48}$ 的熔点 T_m 大，这与图 2.3 的结果是一致的。对于不同形状的 Au-Pt 纳米颗粒，熔点 T_m 遵循 $T_m(Dh) > T_m(Cubo) > T_m(Ih)$ 的顺序。这种

趋势也可以用公式(2.6)来理解,通过计算,可以得到不同形状的形状因子分别为 $\lambda_{Dh} = 0.66$，$\lambda_{Cubo} = 0.95$ 和 $\lambda_{Ih} = 1.40$，也就是 $\lambda_{Dh} < \lambda_{Cubo} < \lambda_{Ih}$。根据数学近似,当 t 非常小时 $\exp(-t) \approx 1 - t$,公式(2.6)可以近似表达为

$$\frac{T_m(x,D,\lambda)}{T_m} \approx 1 - \frac{\lambda S_b(x)}{18RD/D_0(x)} \qquad (2.10)$$

从这个公式中可以明显看出,T_m 随着 λ 的增加而减小。我们知道纳米晶体的熔点 T_m 依赖于其表面体积比。当其他形状纳米颗粒的形状因子 λ 大于球状的时候,其相应的表面体积比也比球状的大。而球状颗粒的表面体积比是随着尺寸的减小而增加的。如果假设具有同样表面体积比的纳米颗粒具有相等的熔化温度,那么相同表面体积比的其他纳米颗粒的熔点 T_m,将等价于更小尺寸的球状纳米颗粒的 T_m。因此,形状因子对 T_m 的影响也是通过纳米粒子的表面体积比来调节的。从公式(2.10)和图 2.5 发现,合金的熔点 T_m 也是随着尺寸减小而降低的,这和单金属纳米颗粒的情况是类似的。此外,H. B. Liu 等(2008)讨论了元素分布对纳米颗粒熔点 T_m 的影响,以及核壳结构、共晶结构和固溶结构的纳米颗粒的稳定性。当温度增加时,这些结构都转化成为最稳定的结构,这意味着不同原子分布的纳米颗粒的熔点 T_m 最后都会归于同一温度范围。由于熔化时经历的过程不同,且元素分布对熔点的影响是有限的,因此这种情况在我们的模型中并没有涉及。

2.2.4　本章小结

基于尺寸和形状影响的熔化温度模型,我们发展了尺寸、成分和形状依赖的熔化温度模型来研究 Au-Pt 合金纳米粒子的熔化现象。结果表明,合金粒子的熔点随着尺寸的减小、成分 Pt 含量 x 的降低和形状因子 λ 的增加而减小。对这个热力学模型可以进一步拓展用来描述类似 Au-Pt 固溶合金的其他双金属化合物的熔化过程。相关的一些分子动力学模拟工作也证实了我们的热力学模型的有效性。

第 3 章 液晶的熔化转变

3.1 液晶熔化的背景简介

理解小尺度空间，尤其是介于介观尺寸范围的低维材料的物理、化学性质，对获得比大块尺寸材料更加优异的性能，并开发新的应用领域都有非常重要的科学意义。我们知道，当尺寸减小到纳米尺度的时候，由于纳米晶体具有大的表面体积比，有机分子晶体熔化和凝固转变的温度将降低，这和金属、半导体的结果是一样的。对于有机分子晶体，分子的位置和位相、自由度都是有序排列在一起的。而在从有序分子晶体到各向异性液体的熔化转变过程中，晶体的有序结构会消失。对于广泛应用于现代光电器件和传感器，以及现代生物材料的液晶材料，拥有介于传统液体和固体之间的性质，其熔化和凝固通常展现出更加复杂的相变过程和行为。当温度提高时，局部位置丢失有序结构导致产生各种不同类型的位相或平移，从而形成有序液晶亚稳相，比如向列液晶相、碟状液晶相和手性液晶相。因而，在相与相之间存在各种不同类型的相变过程。在液晶的应用中，液晶分子经常是被限制在圆柱形纳米孔基体之内的，例如自组装的有序氧化铝纳米孔(AAO)和硅胶。基体纤维状的孔的内壁的限制作用在约束表面层的液晶分子的转动和移动将起到至关重要的作用。此外，不同的表面条件也深刻影响着液晶相变温度的变化。

P-azoxyanisole（PAA）和 4-penty-4'-cyanobiphenyl（5CB）是两种重要的液晶材料，其相变过程经常表现出一定的尺寸依赖性。

对于大块 PAA 和 5CB,当温度增加时,其熔化主要包括两个截然不同的过程:在温度为 T_m^CN 时从晶体相到液晶相(C-N),在温度为 T_m^NI 时从液晶相到液体相(N-I)。当液晶相的尺寸 D(也就是渗入液晶体的氧化铝或硅胶玻璃的小孔洞的尺寸)减小时,相应的相变温度要低于其大块时的转变温度。

最近,根据 Lindemann 规则推导出的纳米晶体原子热振动振幅的尺寸效应模型,一个新的模型提出来描述纳米晶体的熔化温度。这个模型成功描述了有机晶体和半导体的熔化温度,以及聚合物的玻璃转变温度的尺寸效应。本章将以 PAA 和 5CB 为例,通过拓展该熔化模型来描述液晶材料熔化的尺寸效应。这个模型与实验结果拟合得很好。

3.2 液晶熔化的尺寸与界面效应函数理论模型的建立

根据 Lindemann 规则所推导的纳米晶体原子热振动尺寸效应模型,纳米晶体的熔化温度可以表示为

$$\frac{T_\mathrm{m}(D)}{T_\mathrm{m}(\infty)} = \frac{\sigma^2(\infty)}{\sigma^2(D)} = \exp\left[\frac{-(\alpha-1)}{D/D_0 - 1}\right] \quad (3.1)$$

式中,σ^2 表示原子的均方位移。对于分子晶体,D_0 同样可以表示为

$$D_0 = 2(3-d)h \quad (3.2)$$

式中,h 表示分子的直径。

对于自由纳米晶体,将公式(3.1)中的 α 定义为表面的均方位移与晶体内部的均方位移之比,也就是 $\alpha = \sigma_\mathrm{s}^2(D)/\sigma_\mathrm{v}^2(D)$,而 α_s 则可根据下式确定:

$$\alpha_\mathrm{s} = [2S_\mathrm{m}(\infty)/3R] + 1 \quad (3.3)$$

式中,$S_\mathrm{m}(\infty)$ 和 R 分别是晶体的熔化熵和理想气体常数。

我们知道,有机分子晶体的熔化是一个有序—无序转变的过程。而对于 5CB 和 PAA 液晶来说,这个转变过程是分步进行的,即在温度 T_m^CN 时的 C-N 转变和在温度 T_m^NI 时的 N-I 转变。其中,后

一个转变过程主要是类液体—液体的转变,由于这个过程涉及的仅仅是分子形状的变化,其本质控制因素是熵变。在公式(3.3)中,$S_m(\infty)$表示熔化熵。对于液晶纳米晶体来说,$S_m(\infty) = S_m^{CN}(\infty)$,并且$S_m^{NI}(\infty)$是一个附加的 NI 转变熵,其中上角标 CN 和 NI 表示的是对应的转变过程。需要注意的是,$S_m^{CN}(\infty)$和$S_m^{NI}(\infty)$也可以通过实验测量相应的熔化焓$H_m(\infty)$及熔化温度$T_m(\infty)$来间接获得:

$$S_m^{CN}(\infty) = H_m^{CN}(\infty)/T_m^{CN}(\infty) \tag{3.4}$$

$$S_m^{CI}(\infty) = [H_m^{CN}(\infty) + H_m^{NI}(\infty)]/T_m^{NI}(\infty) \tag{3.5}$$

因此,公式(3.3)也可以写成以下形式:

$$\alpha_s^{CN} = [2S_m^{CN}(\infty)/3R] + 1 \tag{3.6}$$

$$\alpha_s^{NI} = [2S_m^{CI}(\infty)/3R] + 1 \tag{3.7}$$

将公式(3.6)和(3.7)代入公式(3.1)中,可以得到

$$\frac{T_m^{CN}(D)}{T_m^{CN}(\infty)} = \exp\left[\frac{-(\alpha_s^{CN} - 1)}{D/D_0 - 1}\right] \tag{3.8}$$

$$\frac{T_m^{NI}(D)}{T_m^{NI}(\infty)} = \exp\left[\frac{-(\alpha_s^{NI} - 1)}{D/D_0 - 1}\right] \tag{3.9}$$

对于被束缚在纳米微孔的液晶体来说,其表面分子与微孔内壁之间的相互作用,比如范德华力和氢键,对相变温度的影响是不能忽略的。此时D_0应该修正为

$$D_0' = 2c(3-d)h \tag{3.10}$$

式中,c是用来规范表面或界面区域的分子势不同于晶体内部的一个常量,对于本节所涉及的处于界面区域的液晶分子,$c = 1/2$;对于纤维线状的微孔系统,$d = 1$。

如果液晶体被约束在惰性环境中,也就是表面分子与微孔内壁之间的界面存在比较微弱的范德华力的时候,此时界面处的化学相互作用与液晶体内比较相似,界面可以近似看成是一个自由表面。把公式(3.10)代入公式(3.8)和(3.9),则有

$$\frac{T_m^{CN}(D)}{T_m^{CN}(\infty)} = \exp\left[\frac{-(\alpha_s^{CN} - 1)}{D/D_0' - 1}\right] \tag{3.11}$$

$$\frac{T_m^{NI}(D)}{T_m^{NI}(\infty)} = \exp\left[\frac{-(\alpha_s^{NI}-1)}{D/D_0'-1}\right] \qquad (3.12)$$

而当液晶分子与微孔内壁之间的界面存在较强的氢键时,氢键对界面处液晶分子的钉扎作用将抑制分子的运动,从而延迟了相变过程,最终将影响到相应的相变温度。这里假设界面处只存在氢键和范德华力两种键合,此时定义氢键数目为 n,总键数为 N,则范德华力的成键数目为 $N-n$,那么氢键的密度可以表示为

$$\rho_H = n/N$$

从而

$$\alpha_i = [(1-\rho_H)\sigma_s^2 + \rho_H \sigma_H^2]/\sigma_v^2$$

式中,σ_H^2 为氢键对液晶分子均方位移的钉扎作用的贡献。因为在液晶内部的分子的运动和在界面处被钉扎的分子的运动都得到抑制,作为近似考虑,我们假设 $\sigma_H^2 = \sigma_v^2$。注意到,当纳米晶体有着自由表面的时候,$\alpha_{max} = \sigma_s^2/\sigma_v^2$,这便是公式(3.11)和(3.12)的情况。因此,

$$\alpha = \alpha_{max}(1-\rho_H) + \rho_H \qquad (3.13)$$

假设所有的液晶表面分子都与微孔内壁发生了键合作用,并且界面处的氢键分布是均匀的,界面处的总 ρ_H 可以近似等于每个分子的氢键密度分布。对于 5CB 分子,只有—C≡N 官能团能与界面处—OH 基团形成氢键。假设其他表面原子,也就是 5CB 分子端点处的氢原子都与内孔壁通过范德华力成键。由于每个 5CB 分子都有 19 个氢原子,所以相应的氢键密度为 $\rho_H = 1/20 = 0.05$;对于 PAA 分子来说,azoxy—官能团则更具有活性而与—OH 基团形成强键合作用。类似地,$\rho_H = 1/15 \approx 0.067$。由于对 5CB 和 PAA,$\rho_H < 1/10$,公式(3.13)可近似简化为

$$\alpha_i = \alpha_{max}(1-\rho_H) \qquad (3.14)$$

把公式(3.14)代入公式(3.11)和(3.12)中,对于界面处有氢键键合的液晶体熔化转变,我们得到

$$\frac{T_m^{CN}(D)}{T_m^{CN}(\infty)} = \exp\left[-\frac{(1-\rho_H)\alpha_s^{CI}-1}{D/D_0'-1}\right] \qquad (3.15)$$

$$\frac{T_{\mathrm{m}}^{\mathrm{NI}}(D)}{T_{\mathrm{m}}^{\mathrm{NI}}(\infty)} = \exp\left[-\frac{(1-\rho_{\mathrm{H}})\alpha_{\mathrm{s}}^{\mathrm{NI}}-1}{D/D'_0-1}\right] \quad (3.16)$$

由于结晶转变是熔化转变的逆过程,类似地,我们也可以定义液晶体凝固温度的尺寸效应为 $T_{\mathrm{f}}(D)/T_{\mathrm{f}}(\infty) = T_{\mathrm{m}}(D)/T_{\mathrm{m}}(\infty)$。也就是,

$$T_{\mathrm{f}}^{\mathrm{NC}}(D)/T_{\mathrm{f}}^{\mathrm{NC}}(\infty) = T_{\mathrm{m}}^{\mathrm{CN}}(D)/T_{\mathrm{m}}^{\mathrm{CN}}(\infty) \quad (3.17)$$

$$T_{\mathrm{f}}^{\mathrm{IN}}(D)/T_{\mathrm{f}}^{\mathrm{IN}}(\infty) = T_{\mathrm{m}}^{\mathrm{NI}}(D)/T_{\mathrm{m}}^{\mathrm{NI}}(\infty) \quad (3.18)$$

表 3.1 是模型计算所需参数。其中液晶分子的平均直径 h,可以依据文献[80]简化计算为 $h = \frac{1}{3}\sum_{i=1}^{3}h_i$,下角标 i 为 1~3 分别代表分子的 x、y 和 z 轴;h_i 是分子沿着相应方向上的长度,平均键长和键角参考文献[217]和[218]。

表 3.1 模型计算所需参数

材料	$T_{\mathrm{m}}^{\mathrm{CN}}$/K	$T_{\mathrm{m}}^{\mathrm{NI}}$/K	$T_{\mathrm{f}}^{\mathrm{NC}}$/K	$T_{\mathrm{f}}^{\mathrm{IN}}$/K	$S_{\mathrm{m}}^{\mathrm{CN}}$/ (J·g−atom·K^{-1})	$S_{\mathrm{m}}^{\mathrm{NI}}$/ (J·g−atom·K^{-1})	h/nm
PAA		408	391		2.376	0.056	0.979
5CB	297	309		309	1.447	0.079	0.989

3.3 理论模型与纳米微孔束缚的 PAA 和 5CB 液晶实验结果的对比与分析

图 3.1 是 5CB 液晶体 C-N 和 N-I 相变温度的尺寸效应。图中虚线、实线、点画线分别代表理论模型公式(3.11),(3.12) 和 (3.16)的预测结果,所需参数在表 3.1 中列出,△、○和□分别是 $T_{\mathrm{m}}^{\mathrm{NI}}$ 的实验结果,◇是束缚在 ODPA(octadecylphosphonic acid) 修饰过的 AAO(nanoporous aluminum oxide) 微孔中的 5CB 的实验结果,☆来自文献[207]。●、▲和■分别是 5CB 的 $T_{\mathrm{m}}^{\mathrm{CN}}$ 的实验结果。

相应地,5CB 的相变温度 $T_{\mathrm{m}}^{\mathrm{CN}}$ 和 $T_{\mathrm{m}}^{\mathrm{NI}}$,PAA 的 $T_{\mathrm{m}}^{\mathrm{NI}}$ 相变温度,以及 PAA 的 $T_{\mathrm{f}}^{\mathrm{NC}}$ 和 5CB 的 $T_{\mathrm{f}}^{\mathrm{IN}}$ 相变温度的实验数据分别如图 3.1,3.2,3.3 所示。

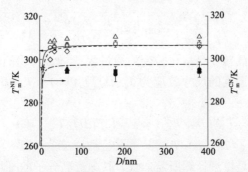

图 3.1　5CB 液晶体 C-N 和 N-I 相变的尺寸效应

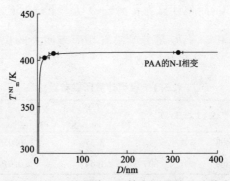

图 3.2　PAA 液晶体 N-I 转变温度的尺寸效应

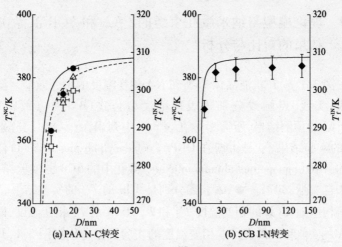

(a) PAA N-C 转变　　　　(b) 5CB I-N 转变

图 3.3　PAA 液晶体的 T_f^{NC} 转变温度的尺寸效应

其中图 3.2 的模型预测是基于公式(3.16)的,相关的参数由表 3.1 给出,实验数据来自于文献[210]。图中虚线和实线分别代表基于公式(3.15)、(3.17)的模型预测和基于公式(3.18)的 5CB 的 T_f^{IN} 转变温度的尺寸效应。图 3.3 a 中带有误差的符号●是 PAA 被束缚在未修饰的凝胶微孔中,Δ 是被束缚在 -Si(CH3)基团修饰过的凝胶微孔中的情况。图 3.3 b 中带误差的符号数据点◆是 5CB 被束缚在未修饰的 CPG 微孔内的实验数据。根据公式(3.11),(3.12),(3.17),(3.18)所预测的结果也在图中给出。

从图 3.1 - 3.3 可以看到,理论预测很好地符合了实验结果。相变温度 T_m^{CN},T_m^{NI},T_f^{CN} 和 T_f^{NI} 都是随着尺寸的减小而下降。尤其当尺寸小于 20 nm 的时候,所有的相变温度都急剧减小。而且,由于 $S_m^{CI} = S_m^{CN} + S^{NI} > S_m^{CN}$,因此 T_m^{CN} 的减小应该弱于 T_m^{NI}。这个结果可以很明显地在图 3.1 中 5CB 被束缚在氢键系统中的情况中观察到。在 $D = 6$ nm 时,$\Delta T_m^{CN}(D) = 8.73$ K,而 $\Delta T_m^{NI}(D) = 9.84$ K,其中 $\Delta T_m(D) = T_m(\infty) - T_m(D)$。基于液晶的相变是熵变导致的这个本质原因,该熔化模型也可以推广到预测和解释其他更为复杂的液晶相变过程。

需要注意的是,在图 3.1 和图 3.3 a 中,对于内孔被 ODPA 修饰过的 5CB 的 T_m^{NI} 转变温度,和被三甲基硅烷(trimethylsilyl)修饰过的由 -Si(CH3)3 基团取代羟基的内孔的 PAA 的 T_f^{NC} 转变温度,都得到了抑制。这是因为受界面间氢键的钉扎作用,液晶体表面处的分子运动受到束缚从而保持有序结构,形状和位相的变化得到延迟。而对于内孔被惰性分子修饰过的界面情况,由于氢键被惰性分子打断,界面对液晶分子的束缚作用变弱,从而使界面处的液晶分子更容易发生运动而产生无序结构。因此,驱动分子运动发生相变而需要的温度 T 就低于界面未经惰性分子修饰过的液晶体。为了进一步简化我们的模型,需要对模型进一步做数学上的近似处理,当 $D \gg D_0(D > 10D_0)$ 时,根据数学公式 $\exp(-x) \approx 1 - x$,公式(3.15)和(3.16)又可以写成

$$\frac{T_m^{NI}(D)}{T_m^{NI}(\infty)} = 1 - \frac{(1-\rho_H)\alpha_s^{NI} - 1}{D/D_0' - 1} \qquad (3.19)$$

$$\frac{T_f^{NC}(D)}{T_f^{NC}(\infty)} = 1 - \frac{(1-\rho_H)\alpha_s^{NC}-1}{D/D_0'-1} \qquad (3.20)$$

可知,T_f^{NC} 和 T_m^{NI} 相变温度对界面处氢键密度 ρ_H 的依赖性可以直接体现出来。ρ_H 越大,相应的相变温度也就越高。此外,由于异质分子的添加也在界面处引入了大量的缺陷结构,这也产生了大量的无序结构和位相缺陷。这些缺陷增加了系统的总熵值,使得破坏液晶表面分子的有序性变得更容易,从而促进了相变过程。

3.4 本章小结

本章通过拓展热力学熔化尺寸效应模型,成功描述了液晶材料的熔化与凝固的尺寸效应现象。结果显示,液晶的熔化和凝固转变随着限制液晶纳米微孔的尺寸减小而下降,这主要是表面体积比的增加导致的。液晶相转变的主要驱动力是不同相变阶段的大块熔化熵。纳米微孔与液晶界面处的环境对液晶相变有很大的影响。界面处的氢键密度是影响相变的主导因素。本章中的模型与相关的实验结果得到了很好的拟合。

第4章 In$_2$O$_3$纳米晶体相变热力学研究

4.1 In$_2$O$_3$纳米晶体相变背景简介

透明半导体氧化物拥有优良的光学透明度和电导,其中 n – 型半导体 In$_2$O$_3$ 的光学能隙为 2.93 eV,因其在太阳能电池中作为光电、微电子设备材料,以及在电极、传感器及平板显示器中的应用而受到广泛关注。在常压情况下,大块 In$_2$O$_3$ 是方铁锰矿类型的体心立方结构(bcc – In$_2$O$_3$,空间群 $Ia\bar{3}$)。高温高压条件将使得 In$_2$O$_3$ 发生相变,转变成另外一个亚稳相结构:斜方六面体刚玉类型结构(rh-In$_2$O$_3$,空间群 $R\bar{3}c$),如图 4.1 所示。rh-In$_2$O$_3$ 结构,尤其是当尺寸减小到纳米尺度时,对稀乙醇及 H$_2$S 都有非常高的敏感度,这

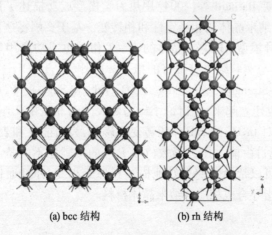

(a) bcc 结构　　(b) rh 结构

图 4.1　In$_2$O$_3$ 的两种结构晶胞示意图

吸引了大量科学家进行实验制备研究纯的及其掺杂的 rh-In_2O_3 纳米晶体。最近,Farvid 等(2010)证明了在常压、温度 523 K 下,rh-In_2O_3 相在小于 5 nm 的范围内存在,而在尺寸大于 5 nm 的时候,相应的结构转变成 bcc-In_2O_3 相。表面能 γ 和表面张力 f 被认为是主要的相变驱动力。另外,在研究常温条件下压力对 In_2O_3 相变的影响时发现,当压力在 15~25 GPa、尺寸约为 6 nm 时,纳米晶体由 bcc-In_2O_3 相转变为 rh-In_2O_3 相,其相变压力与大块时是一样的,当压力释放后两相结构都同时存在。但是另一项研究(Garcia-Domene B,2012)发现,无论大块还是尺寸在 9 nm 时,当压力加到 30 GPa 时都未发现有相变发生。这些研究都暗示着 In_2O_3 的 bcc – rh 转变过程是一个关于温度 T、压力 P 和尺寸 D 的函数。理解这些效应对 In_2O_3 纳米晶体热力学稳定性的影响非常重要,这不仅能让我们可以精确控制合成 In_2O_3 纳米晶体的结构,而且可以拓展它们的潜在应用。

  为了能够更好地理解纳米晶体的相变,到现在为止,一些与实验结果拟合得很好的热力学模型已经相继提出。我们知道,两相之间的热力学稳定性可以通过度量它们各自的吉布斯自由能 G 的值来确定。考虑不同形状纳米晶体的表面、棱和边角处吉布斯自由能的贡献,Barnard 等(2004)用热力学模型成功描述了Ⅳ族半导体和 TiO_2 纳米晶体的相稳定性和相转变。基于亥姆霍兹自由能而建立的德拜模型,Hf 纳米晶体的 D – T 相图也成功应用热力学模型描述,其与实验结果有很好的拟合。此外,结合 BOLS(Bond-Order-Length-Strength)理论,CdSe、Fe_2O_3、CdS 和 SnO_2 的 P,D 与温度 T 的相图也建立起来,并讨论了压力和尺寸对相变的影响。然而,T,P 和 D 对 In_2O_3 纳米晶体相转变的影响,表面能 γ 和表面张力 f 对(吉布斯自由能 – 压力函数(G – P))的贡献还不清楚。而这些函数关系可以帮助我们更好地描述和理解相变,并且通过调制相关参数来指导合成相应的纳米晶体材料。

4.2 In$_2$O$_3$ 纳米晶体相变的温度与压力函数模型的建立

为了满足上述条件,纳米晶体的吉布斯自由能 G 在这里可以写成三个部分的总和,即:表面能的尺寸效应 $G_s(D)$、压力 P 诱导的弹性能的尺寸和压力效应 $G_e(P,D)$、温度 T 诱导的吉布斯自由能的温度效应 $G_v(T,0,\infty)$。据此,可以建立 In$_2$O$_3$ 纳米晶体 bcc-rh 相变的相图,即

$$\Delta G(T,P,D) = \Delta G_v(T,0,\infty) + \Delta G_s(D) + \Delta G_e(P,D) \quad (4.1)$$

$$\Delta G_v(T,0,\infty) = \Delta H - T\Delta S \approx \Delta H[1 - T/T_0(\infty)] \quad (4.2)$$

$$\Delta G_s(D) = \gamma_{rh}A_{rh} - \gamma_{bcc}A_{bcc} = 6(\gamma_{rh}V_{rh} - \gamma_{bcc}V_{bcc})/D \quad (4.3)$$

$$\Delta G_e(P,D) = 4(f_{rh}V_{rh} - f_{bcc}V_{bcc})/D + P_e(V_{rh} - V_{bcc}) \quad (4.4)$$

在上述方程中,Δ 表示的是 rh 和 bcc 两相的差。$\Delta H(\infty)$ 和 $\Delta S(\infty)$ 分别是转变焓和转变熵,它们都是温度 T 的弱函数。$T_0(\infty)$ 为在常压时的大块转变温度。A,V 和 P_e 分别是单位原子的表面积、体积和外压力。γ 为纳米颗粒不同表面表面能的代数平均值,可以表示为

$$\gamma = \frac{\sum_{j=1}^{n} A_j \gamma_j}{\sum_{j=1}^{n} A_j} = \frac{n}{\sum_{j=1}^{n} \frac{1}{\gamma_j}} \quad (4.5)$$

式中,n 为纳米晶粒的小表面数,j 表示第 j 个表面。当忽略液体对表面应力的作用时,可以近似考虑 $f \approx f_{sl}$,f_{sl} 为固液界面应力,因此 f 可以确定为

$$f = (hD_0 S_{vib} H_m B/16V_g R)^{1/2} \quad (4.6)$$

式中,$D_0 = 3h$ 表示球状颗粒的直径,h 为组成化合物的各原子的平均直径;S_{vib} 是熔化熵中的振动熵部分,对于金属氧化物,$S_{vib} \approx S_m(\infty)$。

当系统达到平衡状态,也就是 $\Delta G(T,P,D) = 0$ 时,由尺寸 D 所确定的转变温度 T_c 和转变压力 P_e 的临界点就可以通过下式确定:

$$T_c = T_0 \left[1 - \frac{6(\gamma_{rh}V_{rh} - \gamma_{bcc}V_{bcc}) + 4(f_{rh}V_{rh} - f_{bcc}V_{bcc}) + P_e D(V_{rh} - V_{bcc})}{\Delta HD} \right]$$
(4.7)

$$P_e = \frac{6(\gamma_{rh}V_{rh} - \gamma_{bcc}V_{bcc}) + 4(f_{rh}V_{rh} - f_{bcc}V_{bcc}) + \Delta HD(1 - T/T_0)}{D(V_{bcc} - V_{rh})}$$
(4.8)

因为缺少 rh 相表面能 γ 的数据，而这些数据在公式中是必需的，所以表面能值可通过计算机模拟计算获得。基于第一原理密度泛函理论（DFT）的 Dmol^3 模块，近似方法为 GGA，交互作用势能采用的是 RPBE 方法。所有的堆垛层都被 20 Å 的真空宽度分隔开，以使层间的相互作用最小。表面能定义为 $\gamma = \frac{U_{slab} - U_{bulk}}{2A}$，其中 U_{slab} 和 U_{bulk} 分别表示堆垛层的总能量和相应的大块单元能量，A 为堆垛层一个边的单位表面积。之前的模拟结果和试验数据显示，对于 bcc 相的微观形体结构，由面（111）和边（110）所组成的八面体结构能量最稳定。而 rh 相则通常表现出由一组面（012）面所围成的纳米菱面体结构。相应的示意图见图 4.2。

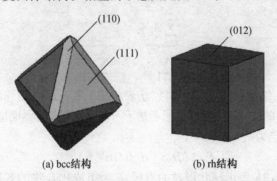

(a) bcc结构　　　　　(b) rh结构

图 4.2　bcc 和 rh 结构理想最稳定的纳米晶微观形貌示意图

这里应该注意的是，对于 rh（012）面，其表面结构较为复杂（如图 4.3 a 所示），我们计算了五种表面结构，包括以 In 和 O 为终端面的所有可能的情况。最稳定的表面结构如图 4.3 b 所示，以 rh01 的计算结果代表公式中用到的表面能。同时计算 bcc 的（111）和

(110)面的表面能用来与以前的计算结果做对比。表面能的计算结果在表 4.1 中给出。相关的表面结构,即 6 层 bcc(111) 和 rh(012) 表面结构及 8 层 bcc(110) 表面结构见图 4.4。注意,一层 rh(012) 面的表面结构在图 4.4 c 中已标示出。表 4.1 中,所计算的表面能结果略大于文献中计算的结果,这主要是由于采用不同的计算方法和模块导致的,但是并不影响模型的预测结果,因为我们只关心 bcc 和 rh 两相的表面能的差值 $\Delta\gamma$。

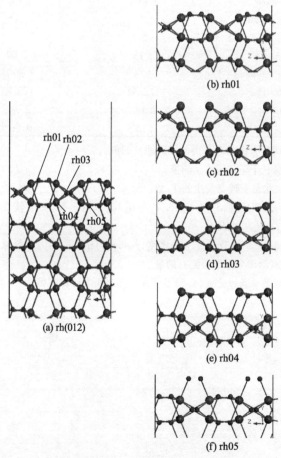

图 4.3 在计算表面能时所考虑的 rh(012) 表面的结构信息

表 4.1 模型计算中的必要参数

	bcc-In$_2$O$_3$	rh-In$_2$O$_3$
$\gamma_{(110)}/(\mathrm{J \cdot m^{-2}})$	1.114[①] 1.070[②]	
$\gamma_{(111)}/(\mathrm{J \cdot m^{-2}})$	1.080[①] 0.890[②]	
$\gamma_{(012)}/(\mathrm{J \cdot m^{-2}})$		0.860[①]
$\rho/(\mathrm{g \cdot cm^{-3}})$	7.18	7.31
T_m/K	2185	
T_0/K	873[③]	
$S_m/(\mathrm{J \cdot g-atom^{-1} \cdot K^{-1}})$	9.61[④]	9.61[④]
$H_m/(\mathrm{kJ \cdot g-atom^{-1}})$	21	26[④]
$\Delta H/(\mathrm{kJ/g-atom})$	5	
$h/\mathrm{\AA}$	2.13	2.07
$V_g/(\mathrm{cm^3 \cdot g-atom^{-1}})$	7.734[⑤]	7.596[⑤]
B/GPa	172.44	172.87
$f/(\mathrm{J \cdot m^{-2}})$	2.146	2.348
$\gamma/(\mathrm{J \cdot m^{-2}})$	1.10[⑥]	0.86[⑥]

注：① 用 DFT 方法计算获得的表面能 γ 的值。

② 文献中计算得到的 γ 的值。

③ 大块状态下转变发生的 T_c 值。

④ 作为近似处理，$S_{mrh} \approx S_{mbcc}$，$T_m = H_m/S_m$。由于 bcc–rh 转变是放热过程，因此 $H_{rh} \approx H_{bcc} + \Delta H$。

⑤ $V = M/\rho$，M 为分子量，ρ 为密度，g-atom 为相应的热力学量除以每个单元原子数，表示每个原子所拥有的量。

⑥ 表面能 γ 通过公式(4.5)及图 4.2 和表 4.1 计算获得。

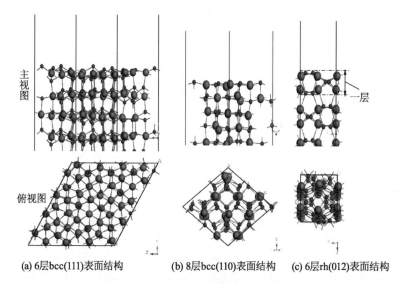

(a) 6层bcc(111)表面结构　　(b) 8层bcc(110)表面结构　　(c) 6层rh(012)表面结构

图4.4　弛豫后的 6 层 bcc-In_2O_3(111)表面结构、8 层(110)表面结构,以及 6 层 rh – In_2O_3(012)表面结构的局部示意图

4.3　相变理论模型和实验结果的对比与分析

图4.5分别描述了根据公式(4.7)和(4.8)计算的在 P_e = 0 GPa,T = 300 K 条件下 In_2O_3 纳米晶体相变的 $T_c(D)$ 函数和 $P_e(D)$ 函数,计算中用到的相关参数已在表4.1中列出。由图可知,T_c 和 P_e 都随着尺寸的减小而单调下降。

在图4.5a中,在外压力 P_e = 0 GPa、温度 T = 523 K 的时候,我们预测到的相变临界直径为 D_c = 3.5 nm,比实验中观察到的结果 D_c = 5 nm 略小。当温度 T 为室温时,D_c = 2.1 nm。图4.5b 所示是在温度 T = 300 K 的时候 bcc – rh 相变的压力随尺寸变化的函数,可以观察到 $P_e(D)$ 明显减小,这与实验结果截然不同。尽管对于大块转变压力,很多实验结果显示在 3.8～30 GPa,但是在压力为 25.24 GPa 时,其晶格常数明显发生变化,而在 D = 6 nm 时,其晶格常数在压力为 15 GPa 时才发生变化。这意味着大块的开始转变压力为 25.24 GPa,而 $D=6$ nm 时为 15 GPa,这支持了模型的预测结果,即 P_e 随 D 的减小而下降。可以清晰地看到,由高压 P 和高温 T 条件下产生的 rh 相为生长 bcc 相结构的"先驱结构",这是由于小尺寸颗粒拥有更大的内压力和表面应力而使得熔化温度降低。在这里,低的熔点降低了 T_m/T 的值,相对于大块材料相当于升高了系统温度。

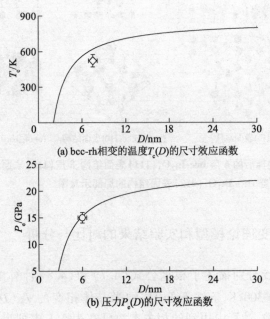

(a) bcc-rh 相变的温度 $T_c(D)$ 的尺寸效应函数

(b) 压力 $P_e(D)$ 的尺寸效应函数

图 4.5　不同条件下纳米晶体相变的 $T_c(D)$ 函数和 $P_e(D)$ 函数

为了阐述各组分对吉布斯自由能差 ΔG 的单独贡献,图 4.6 描绘了 ΔG、ΔG_e、ΔG_s 和 ΔG_v 在不同温度 T 和压力 P 条件下的尺寸效应函数。可见,当尺寸减小时,温度 T 和表面应力 f 都提高了 ΔG,但是表面能 γ 降低了 ΔG。在图 4.6 a 中,转变的临界直径 $D_c = 3.5$ nm,根据公式(4.3)和(4.4)计算得,$\Delta G_e = 1.42$ kJ/g-atom,而 $\Delta G_s = -3.39$ kJ/g-atom。因此,表面能 γ 的贡献要大于 f 而成为 bcc-rh 相变的主要驱动力。在表 4.1 中,我们注意到 $f_{bcc} < f_{rh}$,$\gamma_{bcc} > \gamma_{rh}$,这也证明了 rh 相较低的表面能为诱导 bcc-rh 相变的主因。

在图 4.6 b 中,当温度 T 下降到室温时,D_c = 2.1 nm,由于具体实验中低温、低压时的动力学因素等条件所限,目前还没有相关的实验数据。此时 $|\Delta G_e|$(2.36 kJ/g-atom) < $|\Delta G_s|$(5.64 kJ/g-atom),它们的差值为 3.28 kJ/g-atom。因此 γ 依旧是相变的主要驱动力,此外 ΔG_v 从 2.00 kJ/g-atom 增加到 3.28 kJ/g-atom,这是临界直径 D_c 受温度影响的主要原因。图 4.6 c 为相变压力 P 的影响。当温度 T = 300 K 时,以文献中用到的纳米颗粒直径 D = 6 nm 为参考,根据公式(4.8)计算转变压力为 P_e = 16.2 GPa,这与实验中提到的晶格常数变化的起始压力 15 GPa 非常接近。此时 $|\Delta G_s|$(1.97 kJ/g-atom)和 $|\Delta G_e|$(1.20 kJ/g-atom)的差值为 0.77 kJ/g-atom,已经变得非常小。尽管此时表面能 γ 依旧是相变的主要驱动力,但是它的影响已被强烈削弱,压力 P 的增加降低了 ΔG_e,从而使得表面应力 f 在驱动力的贡献中比例增加。

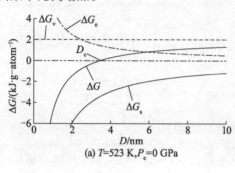

(a) T=523 K, P_e=0 GPa

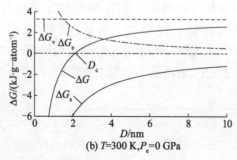

(b) T=300 K, P_e=0 GPa

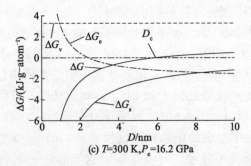

(c) $T=300\ \text{K}, P_e=16.2\ \text{GPa}$

图 4.6 计算出的 ΔG，$\Delta G_v, \Delta G_e$，ΔG_s 在不同温度下的尺寸效应函数

通过上面的讨论知道，表面能 γ 和表面应力 f 是调整 bcc – rh 相变过程的两个主要参数。根据公式(4.7)和(4.8)，可以分别把它们对 T_c 和 P_e 的单独贡献提取出来，也就是

$$T_c^{\gamma} = T_0 \left[1 - \frac{6(\gamma_{\text{rh}} V_{\text{rh}} - \gamma_{\text{bcc}} V_{\text{bcc}})}{\Delta H D} \right] \qquad (4.9)$$

$$T_c^{f} = T_0 \left[1 - \frac{4(f_{\text{rh}} V_{\text{rh}} - f_{\text{bcc}} V_{\text{bcc}})}{\Delta H D} \right] \qquad (4.10)$$

$$P_e^{\gamma} = \frac{6(\gamma_{\text{rh}} V_{\text{rh}} - \gamma_{\text{bcc}} V_{\text{bcc}})}{D(V_{\text{bcc}} - V_{\text{rh}})} \qquad (4.11)$$

$$P_e^{f} = \frac{4(f_{\text{rh}} V_{\text{rh}} - f_{\text{bcc}} V_{\text{bcc}})}{D(V_{\text{bcc}} - V_{\text{rh}})} \qquad (4.12)$$

图 4.7 是根据公式(4.9) – (4.12)画出的随尺寸变化，γ 和 f 对 T_c 与 P_e 的单独贡献函数。显然，随着尺寸的减小，f 同时提高了 T_c^f 和 P_e^f，而 γ 抑制了 T_c^{γ} 和 P_e^{γ}。后者的贡献要大于前者的贡献，最终导致总的结果是 T_c 和 P_e 随尺寸减小而降低。这也证明了在 bcc – rh 相变过程中，表面能 γ 起着主要作用，而表面张力 f 为次要作用。

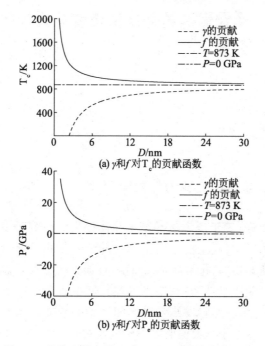

图 4.7　计算出的 γ 及 f 对 T_c 和 P_e 的单独贡献函数

4.4　本章小结

综上所述,通过热力学方法建立了 In_2O_3 纳米晶体 bcc-rh 相变的温度 T 与压力 P 的尺寸效应函数关系。结果表明,转变温度和转变压力都随尺寸的减小而降低,转变压力的变化与实验结果不符。同时详细讨论了 ΔG_e, ΔG_s 和 ΔG_v 对 ΔG 的单独贡献,以及 γ 和 f 对温度和压力的单独贡献。结果证明,rh 相较低的表面能是相变的主要驱动力。随着外压 P 的增加,表面应力对驱动力的贡献越来越大。

第 5 章 确定影响纳米晶体相变压力变化趋势的因素

5.1 纳米晶体相变压力背景简介

近些年来,关于压力诱导纳米晶体相变的研究吸引了越来越多科学家的关注,这主要是由于产生的纳米新相具有与初始相截然不同的新的物理、化学性质,而这些性质在各种工业领域中也可以得到新的应用。然而,在纳米晶体相转变过程中有一个非常值得关注的现象,就是触发纳米晶体相变需要的临界压力 $P_e(D)$ 可以大于也可以小于其大块值 $P_e(\infty)$,而对于这一现象的物理起源现在仍不是很清楚。

众所周知,在压力诱导的相变中,高压亚稳相的密度要大于常压稳定相,也就是它们的原子单位体积遵循关系 $V_L > V_H$,其中 L 与 H 分别表示常压稳定相和高压亚稳相。同样的情况也出现在纳米晶体中。然而,$P_e(D)$ 是尺寸依赖的函数,并且可以大于或者小于 $P_e(\infty)$。对于前者,例如 Fe_2O_3 从 γ 相到 α 相($\gamma-\alpha$),ZnS 从立方闪锌矿到六角纤维锌矿(s-w),CdSe 和 ZnO 从纤维锌矿到岩盐矿(w-r)结构的转变等。但是在一些相变,诸如 CeO_2 从立方萤石结构到正交斜方 α-PbCl2(c-o)结构,AlN 从纤维锌矿到岩盐矿结构(w-r),ZrO_2 从单斜晶系结构到四角晶系结构(m-t),Fe 和 $Fe_{30}Cu_{10}$ 从 bcc 到 hcp 结构的转变中,发现 $P_e(D)$ 是减小的。对 $P_e(D)$ 的抑制作用主要归因于尺寸诱导的体膨胀以及泊松比和剪切模量的软化。此外,在 In_2O_3 从立方方铁锰矿到六角晶系金刚砂结构(bcc-rh),SnO_2 从金红石到萤石结构的转变中发现,$P_e(D) = $

$P_e(\infty)$。但是最近的研究同样发现了 $P_e(D)$ 在 In_2O_3 和 SnO_2 的相变中分别是减小和增加的。这些多样性结果经常让实验学家和理论工作者们感到困惑。因此探究这些相变差异的起源及建立一个标准来描述这些现象就显得非常有意义。

热力学理论可以精确描述在压力诱导相变过程中能量的转移及变化。一些热力学模型已经提出来描述这种纳米晶体相变的尺寸依赖性。由于随着纳米晶体尺寸 D 的减小,表面体积比变得非常大,此时表面应力 f 和表面能 γ 对相变的贡献就变得非常重要。另外,压力对纳米晶相变的贡献又被分为三个部分,即体积坍塌比,两相表面能的差和大块内能差,但是 f 对 $P_e(D)$ 的贡献被忽略了。需要注意的是,γ 描述的是形成固体表面单位面积上需要的可逆功,而 f 表示的是由于单位面积上的弹性变形而导致的可逆功,等于 γ 对表面应变的切线方向求导。尽管上述实验和理论结果丰富了我们对 $P_e(D)$ 函数的理解,但是仍需要理论模型来定量评估确定 $P_e(D)$。

在本章中,根据以前提出的纳米晶相变热力学模型,针对 γ 和 f 对 $P_e(D)$ 的影响进行深入的讨论,并提出一个简单方法来判断 $P_e(D)/P_e(\infty)$ 的变化趋势。这些结果可以帮助我们更好地理解为什么纳米晶体的临界转变压力会出现增加或减小的不同趋势。

5.2 确定纳米晶体相变压力变化趋势的理论方法

在参量 P、T(开尔文温度)、D 都确定的条件下,系统中拥有最小吉布斯自由能(G)的相结构是最稳定的。当系统中的两相达到平衡状态时,它们之间的吉布斯自由能差消失,也就是 $\Delta G = 0$。在温度 T 恒定时,P 和 D 是 ΔG 的热力学变量,可以考虑 ΔG 由三部分组成,即大块吉布斯自由能差 $\Delta G(\infty)$、尺寸依赖的表面能差 $\Delta G_s(P,D)$ 和尺寸与压力依赖的弹性能差 $\Delta G_e(P,D)$:

$$\Delta G(P,D) = \Delta G(\infty) + \Delta G_s(D) + \Delta G_e(P,D) \quad (5.1)$$

其中 $\Delta G(\infty)$ 可以表示成

$$\Delta G(\infty) = -P_e(\infty)\Delta V \tag{5.2}$$

这里 $\Delta V = V_H - V_L$。$\Delta G_s(D)$ 可以通过 γ 和 V_g 推导得到:

$$\Delta G_s(D) = 6(\gamma_H V_H - \gamma_L V_L)/D \tag{5.3}$$

函数 $\Delta G_e(P,D)$ 主要受 P_{in}（由 f 引起的内压力）和 P_e 影响,即 $P = P_{in} + P_e$。当 $P_e \approx 0$ 时, $P = P_{in}$。根据 Laplace-Young 方程,有

$$P_{in} = 4f/D \tag{5.4}$$

式中 f 可以近似考虑为固—液表面应力 f_{sl},

$$f_{sl} = (hD_0 S_{vib} H_m B/16VR)^{1/2} \tag{5.5}$$

式中,h 为原子直径,B 为大块模量,对于理想球体粒子 $D_0 = 3h$,R 为理想气体常量,H_m 为在 T_m 下原子单位的大块熔化焓,S_{vib} 为熔化熵中的振动部分,并且对于有机物和半导体而言 $S_{vib} \approx S_m$。考虑公式 (5.4) 和 ΔG_e 的定义,有

$$\Delta G_e(P,D) = P_H V_H - P_L V_L = 4(f_H V_H - f_L V_L)/D + P_e(D)\Delta V \tag{5.6}$$

在平衡状态,$\Delta G(P,D) = 0$。根据公式 (5.1),(5.2),(5.3) 和 (5.6),有

$$P_e(D) = P_e(\infty) - \frac{6(\gamma_H V_H - \gamma_L V_L)}{D\Delta V} - \frac{4(f_H V_H - f_L V_L)}{D\Delta V} \tag{5.7}$$

或

$$\frac{P_e(D)}{P_e(\infty)} = 1 - \frac{6(\gamma_H V_H - \gamma_L V_L)}{DP_e(\infty)\Delta V} - \frac{4(f_H V_H - f_L V_L)}{DP_e(\infty)\Delta V} \tag{5.8}$$

由公式 (5.7),如果 $\Delta P_e = P_e(D) - P_e(\infty) > 0$,我们定义 $\Delta(\gamma V) = \gamma_H V_H - \gamma_L V_L$ 和 $\Delta(fV) = f_H V_H - f_L V_L$,则

$$3\Delta(\gamma V) + 2\Delta(fV) > 0 \tag{5.9}$$

相反,如果 $\Delta P_e < 0$,则

$$3\Delta(\gamma V) + 2\Delta(fV) < 0 \tag{5.10}$$

注意,公式 (5.7) 也可以写成

$$\Delta P_e D = -\frac{[\Delta 6(\gamma V) + \Delta 4(fV)]}{\Delta V} = \kappa \tag{5.11}$$

公式 (5.11) 右边部分是不依赖尺寸的,对于确定的两相系统。κ 为常量,表示 $P_e(D)$ 对 D 增加或减小的依赖性的强弱趋势。

为了阐述 γ 和 f 对纳米晶体相变的单独贡献,根据公式 (5.8),

它们的尺寸效应可以确定为

$$\frac{P_e^\gamma(D)}{P_e(\infty)} = -\frac{6\Delta(\gamma V)}{DP_e(\infty)\Delta V} \quad (5.12)$$

$$\frac{P_e^f(D)}{P_e(\infty)} = -\frac{4\Delta(fV)}{DP_e(\infty)\Delta V} \quad (5.13)$$

表5.1为上述计算过程中用到的一些参数。

表5.1 计算过程中用到的参数

	γ/ (J·m^{-2})	f/ (J·m^{-2})	V_g/ (cm^3·g-atom^{-1})	$\Delta(\gamma V)$	$\Delta(fV)$	$-\Delta V$/ (cm^3·g-atom^{-1})
w-CdSe	0.48	0.54	16.90	0.854	2.362	2.89
r-CdSe	0.64	0.82	14.01			
w-ZnO	0.50	0.69	7.17	8.215	2.132	1.27
r-ZnO	2.00	1.20	5.90			
w-ZnS	0.57	0.94	12.1	3.337	4.215	0.2
s-ZnS	0.86	1.31	11.9			
γ-Fe$_2$O$_3$	0.96	5.67	6.60	2.176	3.557	0.52
α-Fe$_2$O$_3$	1.40	6.74	6.08			
m-ZrO$_2$	1.70	5.30	7.39	-2.333	-0.738	0.66
r-ZrO$_2$	1.52	5.71	6.73			
bcc-In$_2$O$_3$	1.10	2.16	7.73	-1.976	1.315	0.13
rh-In$_2$O$_3$	0.86	2.37	7.60			
γ-Al$_2$O$_3$	1.50	4.49	5.56	1.88	-0.002	0.45
α-Al$_2$O$_3$	2.00	5.57	5.11			
α-TiO$_2$	1.34	4.28	7.01	1.330	1.845	0.55
b-TiO$_2$	1.66	4.93	6.46			
γ-TiO$_2$	1.93	5.18	6.26	1.358	0.579	0.2

5.3 理论方法和实验结果的对比与分析

图5.1展示的是根据公式(5.9),(5.10),(5.11)对一些半导体和金属氧化物的预测结果,实线表示$3\Delta(\gamma V)+2\Delta(fV)=0$的情况,此时$\Delta P_e(D)=0$。实心符号是通过公式(5.9)计算得到的结果,对应于左坐标;空心符号根据公式(5.10)得到,对应的是右边的坐标。实心符号处于实线以上的部分表示的是$\Delta P_e(D)>0$,满

足公式(5.9)。相反,满足公式(5.10)的计算结果都处于该条线以下。这些预测都与实验结果及理论计算符合得很好。需要注意的是,尽管有实验报道对 In_2O_3 有 $\Delta P_e(D) = 0$,但最近的计算结果显示 $\Delta P_e(D)$ 是略小于 0 的。此外,我们同样预测了 γ-Al_2O_3 和 α-TiO_2 纳米晶体的 $\Delta P_e(D)$ 是增加的。

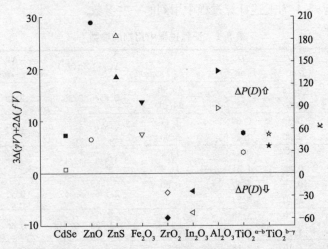

图 5.1 由两相的 γ, f 和 V_g 确定的 $P_e(D)$ 随尺寸 D 的减小变化的趋势,以及基于所确立的 κ 描述 ΔP_e 对 D 的依赖强度关系

根据公式(5.9),在常温常压下 $f = \partial G/\partial A = \gamma + A\partial\gamma/\partial A$,其中 A 为表面积。因此,表面能 γ 是影响 ΔP_e 增加或减小的主导因素。在一般情况下,$\partial\gamma/\partial A > 0$,因此 $f > \gamma$ 且二者都为正值。这就是公式(5.9)和(5.10)中系数分别为 3 和 2 的原因。由公式(5.8)知,当 $\gamma_H > \gamma_L$ 时,将会使 ΔP_e 的值增加。然而,如果 $\gamma_L > \gamma_H$ 满足公式(5.10),那么 ΔP_e 相反的趋势就会呈现出来,如表 5.1 中 ZrO_2 和 In_2O_3 所示。同时,由于 $\partial G = V\partial P$,所以有 $f = V\partial P/\partial A$。因此对于大块结构,压力 P 的增加将导致 f 的提高,从而可以推断 $f_L < f_H$。当 $f_L < f_H$ 时,由公式(5.9)知,将直接产生大的 $\Delta(fV)$ 值,从而对 $\Delta(\gamma V)$ 值的要求减小,这样,对一些 $\gamma_L > \gamma_H$ 时的情况也是可以接受的。注意,由于压力诱导相变导致 $\Delta V < 0$。

对应于图 5.1 右坐标,空心符号代表的是由公式(5.11)计算出的结果 κ,其意义是 ΔP_e 对 D 依赖的强度。大的 $|\kappa|$ 值表现了 ΔP_e 对 D 具有较强的依赖性,反之,依赖性就弱。在列出的半导体材料中,可以清晰地发现 CdSe 拥有最小 κ 值,而 ZnS 的 κ 值最大,这可以从最近的理论和实验结果(S. Li,2006;S. Li,2008;H. Z. Zhang,2003;R. S. Kumar,2007)中得到证实。当尺寸 D 从 10 nm 减小到 5 nm 时,CdSe 的 ΔP_e 从 2.98 GPa 增加到 3.46 GPa,其变化差为 0.48 GPa。而对于 ZnS,ΔP_e 则从 10.09 GPa 增加到 20.17 GPa,其变化差为 10.08 GPa。根据公式(5.11),我们同样可以判断 γ 和 f 对 ΔP_e 对 D 依赖强度的影响。κ 值直接由 $\Delta(\gamma V)$ 和 $\Delta(fV)$ 来决定,这意味着较低的 γ_L 和 f_L 或较高的 γ_H 和 f_H 有利于加速 ΔP_e 的变化。

尽管依据公式(5.9),(5.10) 和 (5.11) 可以确定 $\Delta P_e(D)$ 的变化趋势,但是对于 γ,f 和 V_g 对 $P_e(D)$ 的单独贡献的细节信息依然不够清楚,而公式(5.12),(5.13) 和图 5.2 则对此进行了详细的描述。图 5.2 所示为几种不同情况的 $P_e(D)$ 函数:在图 5.2 a 中,γ 和 f 都提高了 CdSe 的 $P_e(D)$,而 f 对 $P_e(D)$ 的影响要大于 γ;图 5.2 b 所示的是 ZrO_2 纳米晶体,当尺寸 D 减小时,$P_e(D) < P_e(\infty)$ 而 γ 的作用要明显强于 f;最有意思的情况是图 5.2 c 中的 In_2O_3 纳米晶体,随着尺寸 D 的减小,尤其是当 $D < 2$ nm 时,γ 和 f 对 $P_e(D)$ 的贡献表现出了不同的趋势,即 f 增强了 $P_e(D)$ 而 γ 减弱了 $P_e(D)$,且 γ 对 $P_e(D)$ 的削弱影响要大于 f 的增强的影响,γ 和 f 的共同作用导致 $P_e(D) < P_e(\infty)$。对于 $P_e(\infty)$ 值,CdSe 引用 2.5 GPa,ZrO_2 引用 2.9 GPa,In_2O_3 引用 20 GPa,后二者分别取其均值 2.5~3.3 GPa 和 15~25 GPa。观察公式(5.12)和(5.13),由于 $\Delta V = V_H - V_L < 0$,所以 $P_e^\gamma(D)$ 和 $P_e^f(D)$ 的符号直接取决于 $\Delta(\gamma V)$ 和 $\Delta(fV)$,但是 γ 和 f 的单独贡献并不能确定 $P_e^\gamma(D)$ 和 $P_e^f(D)$ 的变化趋势。γ 和 f 对 $\Delta P_e(D)$ 做出的正(负)的贡献的必要条件应该是对于 $\Delta P_e^\gamma(D)$ 为 $\gamma_H/\gamma_L > V_L/V_H$ 和对于 $\Delta P_e^f(D)$ 为 $f_H/f_L > V_L/V_H$(或 $\gamma_H/\gamma_L < V_L/V_H$ 和 $f_H/f_L < V_L/V_H$)。$\Delta P_e^\gamma(D)$ 和 $\Delta P_e^f(D)$

的共同作用确定了 $\Delta P_e(D)$ 在纳米晶体相变的变化方向。在图 5.2 中,我们发现对于 ZrO_2 和 In_2O_3,γ 是改变 $\Delta P_e(D)$ 的主要因素,而对于 CdSe,f 起主导作用。对于这一点我们也可以通过比较 $-3\Delta(\gamma V)$ 和 $-2\Delta(fV)$ 来确定,其相对值较大的量确定了相应参数 (γ 或 f) 对 $\Delta P_e(D)$ 的主导作用。

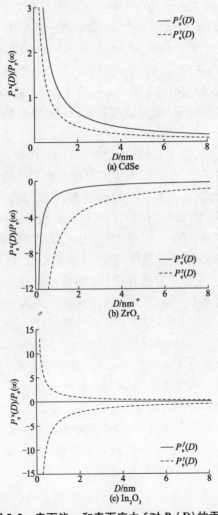

图 5.2 表面能 γ 和表面应力 f 对 $P_e(D)$ 的贡献

5.4 本章小结

综上所述,我们发展出了一套可以预测纳米晶体相变的临界压力变化趋势的模型。结果表明,临界压力 $P_e(D)$ 的变化趋势及其对尺寸 D 的依赖强度可以由两相 γ, f 和 V_g 的值来确定。此外,表面能 γ 和表面应力 f 对 $P_e(D)$ 的单独贡献,即 $P_e^{\gamma}(D)$ 和 $P_e^{f}(D)$,本章也进行了详细的讨论,预测结果与相应的实验结果及理论研究一致。

第6章 尺寸和表面配位因素对纳米晶体表面能的影响

6.1 纳米晶体表面能背景简介

随着尺寸的减小,纳米结构材料往往表现出与块体材料截然不同的物理、化学性质,因此引起了广泛的研究兴趣。表面能 γ 是非常重要的一个物理量,对于理解材料的表面重构、表面弛豫、晶体生长、表面熔化等现象有重要意义。由于随着纳米晶体尺寸减小,表面体积比急剧增加,其他一些热力学量,如结合能、潜热、纳米晶体的熔化温度及过热温度等都可以关联到表面效应上。因此,描述表面能的尺寸效应函数 $\gamma(D)$ 对理解纳米尺度材料表现出的一些行为至关重要。

迄今为止,针对 $\gamma(D)$ 函数已经提出了一些理论模型。Tolman 根据毛细管理论发展了对应的理论模型,发现在更宽的尺寸范围内,随着尺寸的减小,$\gamma(D)$ 是下降的。此外,还讨论了小团簇和纳米晶体的形状因素对 $\gamma(D)$ 的影响。H. M. Lu 等(2008)发现 $\gamma(D)$ 也是随 D 减小而下降的,但是纳米晶体每个晶面的 $\gamma(D)$ 值比率是不依赖于尺寸的。$\gamma(D)$ 类似的尺寸依赖性关系也得到了 BOLS 模型、液滴模型、连续介质理论,以及 Aqra、Aïssa 和 Zhang 等人的理论工作的佐证。随着尺寸 D 的减小,纳米晶体表面原子键缺失,从而使表层原子出现表面重构和弛豫增强,这被认为是导致 $\gamma(D)$ 发生变化的根本原因。然而,当 Nanda(2003)研究 Ag 纳米粒子蒸发温度的时候,观察到了截然相反的现象。与 Ag 的块体表面能值 $1.2 \sim 1.4$ J/m^2 相比,出现了较高的 $\gamma(D)$ 值 7.2 J/m^2,这与从晶格收缩实验中外推得到的结果相近（$\gamma = 6.4 \ \text{J/m}^2$）。这种反常现象可以由面心立方晶体（FCC）金属晶格收缩时产生的负的化学表面能系数来解释。Medasani 在利用第一性原理和嵌入原子势方法分别计算半径 $r < 1$ nm 和 $r < 50$ nm 的 Ag 纳米粒子的时候，也得到了随尺寸的下降 $\gamma(D)$ 反而增加的结果。此外，单位体积表面能密度、单位原子残余结合能等概念也相继被提出以重新定义表面能。F. M. Takrori（2017）的相关报道讨论了二元合金的表面能函数 $\gamma(D)$。这些理论模型都是基于不同的假设，而由失配表面原子所导致的表面键收缩和表面弛豫被认为是主导因素。表面断键数 Z_{b-s} 描述表面失配原子个数，是估算纳米晶体表面能 $\gamma(D)$ 的一个重要的物理量，其中 Z 为配位数，b 和 s 分别代表块体和表面。$\gamma(D)$ 与 Z_{b-s} 之间的关联性目前仍不清楚，因此，需要发展对应新的理论模型来阐述这种关系。

6.2　纳米晶体表面能的尺寸与表面配位函数模型的建立

考虑体系内能的变化和 Mott 公式中熔化熵中振动熵的表达式，纳米晶体的熔化温度推导如下：

$$\frac{T_{\mathrm{m}}(D)}{T_{\mathrm{m}}(\infty)} = \frac{1}{2} - \frac{S_{\mathrm{m}}(\infty)}{3R} + \sqrt{\left[\frac{1}{2} - \frac{S_{\mathrm{m}}(\infty)}{3R}\right] + 2\left[\frac{S_{\mathrm{m}}(\infty)}{3R} - \frac{Z_{b-s}(E_c - E_l)D_0}{3DRT_{\mathrm{m}}(\infty)}\right]} \tag{6.1}$$

式中，$T_{\mathrm{m}}(\infty)$，$S_{\mathrm{m}}(\infty)$，R，E 分别为块体的熔化温度、熔化熵、理想气体常数和每摩尔结合能；下标 c 和 l 分别代表晶体和液体；D_0 为所有原子都位于表面处时的临界尺寸，可由维度 d 和原子直径 h 来计算：$D_0 = 2(3-d)h$，其中 $d = 0,1,2$ 分别对应纳米粒子、纳米线和纳米薄膜。

表面能可根据公式 $\gamma(\infty) = kE_c(\infty)$ 直接和结合能 $E_c(\infty)$ 相关联，其中 $k = 1 - Z_s/Z_b$ 为不依赖于尺寸但决定于配位数的材料常量。当发生晶体结构转变,在键的性质不改变时,转变能要远小于结合能,所以作为一级近似,假设结合能 $E_c(\infty)$ 不依赖于晶体结构。$E_c(\infty)$ 反映键强的性质,而 $\gamma(\infty)$ 又直接由表面原子键合所决定,此外 $E_c(\infty)$ 又直接与 $T_m(\infty)$ 成比例关系,即 $E_c(\infty) = \frac{3k_B e T_m(\infty)}{nf^2}$,其中 k_B, e, n 及 f 分别为玻尔兹曼常量、原子价、原子间斥力指数和原子在 $T_m(\infty)$ 的位移与平衡原子间距之比。因此,当纳米晶体拥有与块体相同晶体结构时,$\gamma_s(\infty) \propto E_c(\infty) \propto T_m(\infty)$。此关系拓展到纳米尺度,并结合公式(6.1),则有

$$\frac{\gamma_s(D)}{\gamma_s(\infty)} = \frac{E_c(D)}{E_c(\infty)} = \frac{T_m(D)}{T_m(\infty)}$$

$$= \frac{1}{2} = \sqrt{\left[1 - \frac{S_m(\infty)}{3R}\right] + 2\left[\frac{S_m(\infty)}{3R} - \frac{Z_{b-s}(E_c - E_1)D_0}{3DRT_m(\infty)}\right]}$$

(6.2)

表 6.1 为计算过程中所需参数。

表 6.1 计算过程中所需参数

	T_m/K	S_m^a/(J·mol^{-1}·K^{-1})	H_m/(kJ·mol^{-1})	U_c/(kJ·mol^{-1})	U_L/(kJ·mol^{-1})	h/nm	γ_s/(J·m^{-2})
Au	1337.3	9.5	12.7	364.0	343.1	0.288	1.36
Au (111)							
Au (100)							1.63
Au (110)							1.70
Ag	1234.9	9.2	11.3	284.5	257.7	0.289	1.06
Al	933.5	11.5	10.7	322.2	290.8	0.286	1.03
Al (110)							1.30
Cu	1357.8	9.6	13.0	338.9	306.7	0.282	1.59
Be (0001)	1560.0	6.3	9.8	326.4	308.8	0.222	1.63
Na (110)	370.9	7.0	2.6	108.4	99.2	0.372	0.26
Ni (110)	1728	10.2	17.6	422.6	374.8	0.270	2.45
Mg (0001)	923	9.8	9.04	148.5	127.6	0.320	0.79
Si	1687	23.5	39.6	439.3	383.3	0.1568	1.24

6.3 纳米晶体表面能模型和金属、半导体、合金实验结果的对比与分析

图 6.1 为不同理论模型对 Au,Ag 和 Al 纳米粒子的 $\gamma(D)$ 函数预测及与相应的实验数据对比。其中实线是公式 6.2 预测的结果,虚线、点画线和双点画线分别是其他理论预测的结果。图中,▽引自文献[287];○为由结合能转换得到的表面能值,引自文献[288];◇引自文献[287];+ 与 ☆ 分别引自文献[289],[290]。假设面心立方结构的平均断键数为 7。很明显,$\gamma(D)/\gamma(\infty)$ 随着尺寸 D 的下降而减小,这与实验结果和其他理论模型预测一致。但是根据公式(6.2),$\gamma(D)/\gamma(\infty)$ 与尺寸满足关系 $\gamma(D)/\gamma(\infty) \propto (C-1/D)^{1/2}$,这与其他模型建立的 $\gamma(D)/\gamma(\infty) \propto C'-1/D$ 关系有所不同,其中 C 与 C' 分别为常量。

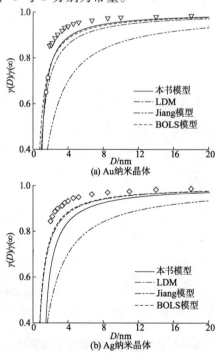

(a) Au 纳米晶体

(b) Ag 纳米晶体

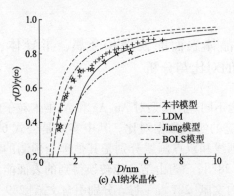
(c) Al纳米晶体

图 6.1 理论模型预测与实验结果对比

当负曲率表面存在时,负曲率表面上的断键将直接影响其表面能和表面原子稳定性。图 6.2 给出了 Cu 纳米粒子和纳米孔的表面能 $\gamma(D)/\gamma(\infty)$ 的尺寸函数。

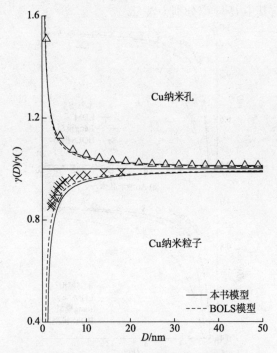

图 6.2 Cu 纳米粒子和纳米孔表面能

图中实线根据公式(6.2)绘出,虚线根据 BOLS 模型预测的结果绘制。×与▽分别引自文献[287]和[291],△为文献[292]中预测的 Cu 纳米孔的表面能。需要注意的是,对于负曲率表面,用公式(6.2)描述纳米孔时其尺寸由 $-D$ 取代。因为当纳米粒子尺寸减小时,表面出现键收缩,表面原子稳定性增强,因此 $\gamma(D)$ 下降。而此处的负号则意味着随着尺寸的减小,纳米孔表面将出现键拉伸,从而使负曲率表面上的原子稳定性下降,导致 $\gamma(D)$ 增加。在图 6.2 中很容易观察到这种趋势,对纳米孔和纳米粒子来说,$\gamma(D)/\gamma(\infty)$ 的增加和减小以 $\gamma(D)/\gamma(\infty)=1$ 呈对称分布,这与 BLOS 模型的预测相吻合。这意味着断键效应导致的纳米孔负曲率表面原子不稳定性与纳米粒子表面原子的稳定性是等价的。这也是小尺寸纳米孔和纳米粒子拥有更大驱动力发生聚集的原因所在。

图 6.3 为一些金属纳米晶体不同低指数表面的理论模型与实验数据之间的对比,包括 Be(0001),Mg(0001),Al(110),Ni(110),Na(110) 等,其中 Be(0001)(图中○)与 Mg(0001)(图中□)引自文献[293],Al(110)(图中▲)引自文献[294],Ni(110)(图中●)引自文献[295],Na(110)(图中■)引自文献[296]。

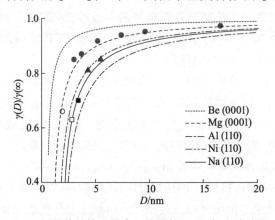

图 6.3　不同晶体结构金属表面能函数 $\gamma(D)$ 与实验结果对比

在这些金属元素中,Al 和 Ni 为面心立方 FCC 结构,Na 为体心立方(BCC)结构，Be 和 Mg 为密排六方(HCP)结构。晶体结构对表面配位有非常重要的影响，也直接影响表面原子的键强和稳定性。尤其对 BCC 和 HCP 结构，最近邻和次近邻原子都对配位数做贡献。因此，不同晶体结构的失配数 Z_{b-s} 可由下面的公式得出：

FCC 晶体结构：

$$Z_{b-s} = Z_b - Z_{(hkl)} = \begin{cases} 2h+k, & h, k, l \text{ 都为奇数} \\ 4h+2k, & \text{其他} \end{cases} \quad (6.3)$$

BCC 晶体结构：

$$Z_{b-s} = Z_b - Z_{(hkl)} = \begin{cases} 2h+(h+k+l), & h+k+l \text{ 为偶数} \\ 4h+2(h+k+l), & h+k+l \text{ 为奇数且 } h-k-l \geqslant 0 \\ 2(h+k+l)+2(h+k+l), & h+k+l \text{ 为奇数且 } h-k-l < 0 \end{cases} \quad (6.4)$$

HCP 晶体结构：

$$Z_{b-s} = Z_b - Z_{(hkil)} = \begin{cases} 4(h+2k)+3l, & \text{针对}(0001)\text{面} \\ 4(h+k)+(8h+4k)/3, & \text{针对}(10\bar{1}0)\text{面} \end{cases} \quad (6.5)$$

因此，对于 Al(110) 和 Ni(110) 面，$Z_{b-s} = 6$；对于 Na(110)，$Z_{b-s} = 4$；对于 Be(0001) 和 Mg(0001)，$Z_{b-s} = 3$。显然，图 6.3 中预测的 $\gamma(D)$ 值与实验结果具有很好的一致性，尽管对于 FCC 结构，Al(110) 和 Ni(110) 的 $\gamma(D)$ 略微低估；对于 HCP 结构，Be(0001) 和 Mg(0001) 的 $\gamma(D)$ 略微高估。

在公式(6.2)中，对于以 D 为确定尺寸的纳米晶体，表面能可写成 γ_D。γ_D 与表面断键数 Z_{b-s} 之间的关系可表示为 $\gamma_D(Z_{b-s})/\gamma(\infty) \propto (K - Z_{b-s}/Z_b)^{1/2}$，其中 K 为常量。可以看出，减小纳米晶体表面上的 Z_{b-s} 将增加对应的 $\gamma_D(Z_{b-s})$ 或减缓 $\gamma(D)/\gamma(\infty)$ 的下降，这种趋势在图 6.4 中给出。图中，空心符号◇，○，□分别为引用文献[297]中(111)，(100)及(110)面的数据，对应实心符号为模型预测的结果。由图可知，3.8 nm 的 Au 纳米晶体的表面能 $\gamma(D)$ 值分别为 $\gamma_{3.8}^{(111)} = 0.928$ J/m^2，$\gamma_{3.8}^{(100)} = 1.148$ J/m^2 及 $\gamma_{3.8}^{(110)} = 1.184$ J/m^2，而相应的块体值分别为 $\gamma(111) = 1.28$ J/m^2，$\gamma(100) =$

1.63 J/m² 和 γ(110) = 1.70 J/m²。也就是说,它们满足关系 $[\gamma_{3.8}^{(111)}/\gamma(111) = 0.725] > [\gamma_{3.8}^{(100)}/\gamma(100) = 0.704] > [\gamma_{3.8}^{(110)}/\gamma(110) = 0.696]$,这与公式(6.2)中预测的结果 0.915 > 0.886 > 0.851 相吻合。即当尺寸 D 减小时,$\gamma(D)$ 的下降满足如下顺序: $[\Delta\gamma(110) = 0.516 \text{ J/m}^2] > [\Delta\gamma(100) = 0.482 \text{ J/m}^2] > [\Delta\gamma(111) = 0.352 \text{ J/m}^2]$,其中 $\Delta\gamma = \gamma(\infty) - \gamma(D)$。这种趋势可以由 $\gamma(D)/\gamma(\infty)$ 和 Z_{b-s} 的函数关系进行解释。对于 Au,有 $[Z_{b-s}^{(111)} = 3] < [Z_{b-s}^{(100)} = 4] < [Z_{b-s}^{(110)} = 5]$,其中(110)面拥有最多的断键数,而(111)面的断键数最少,因此,(110)拥有最低的表面稳定性或最高的表面能,当 D 减小时,将导致最强的表面键收缩。(111)面的情况则完全相反。

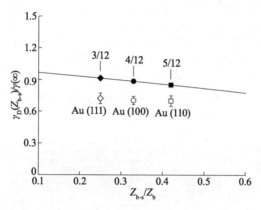

图 6.4 尺寸为 3.8 nm 的 Au 纳米粒子表面能与表面断键数之间的关系

图 6.5 比较了公式(6.2)和文献[298]中针对 Si 纳米孔的 $\gamma(D)$ 值的预测结果。对于 Si(110)面,Z_s 值的最近邻和次近邻配位数分别为 3 和 6,而 Z_b 最近邻和次近邻配位数为 4 和 12。因此,Z_{b-s} 可近似为 7/16。公式(6.2)对 Si 纳米孔的 $\gamma(D)$ 的预测结果与其他理论吻合。图 6.6 为对 Cu-Al 和 Cu-Au 合金的预测。此时假设二元合金皆为规则固溶体,因此公式(6.2)中二元合金的熔化焓 $H_m(x, \infty)$,$T_m(x, \infty)$,$h(x, \infty)$,$E_c(x, \infty)$ 和 $E_1(x, \infty)$ 可以近似用公式 $M(x, \infty) = xM(A, \infty) + (1-x)M(B, \infty)$ 来计算,其中 M

代表对应的物理量，x 代表合金中成分"A"的摩尔分数。$S_m(x,\infty)$ 可由关系 $S_m(x,\infty)=H_m(x,\infty)/T_m(x,\infty)$ 进行确定。对于 Cu-Al 及 Cu-Au 合金，计算得到的 $\gamma(x,D)$ 也是随尺寸 D 下降而减小的，这与 Takrori 的理论模型预测相吻合。图 6.5 和图 6.6 表明，公式 (6.2) 也可以用来描述半导体和二元合金纳米晶体的表面能。

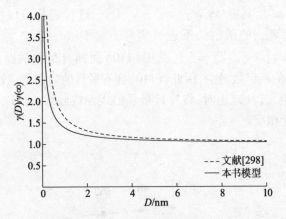

图 6.5　针对 Si 纳米孔的 $\gamma(D)$ 计算结果对比

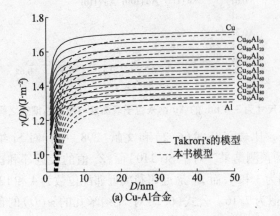

(a) Cu-Al 合金

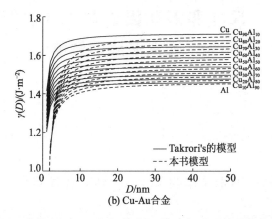

图 6.6 针对 Cu-Al 合金和 Cu-Au 合金 $\gamma(D)$ 的预测结果对比

6.4 本章小结

基于熔化温度的尺寸效应函数,本章提出了无任何可调参数的解析模型来描述纳米晶体和纳米孔的尺寸和断键数对表面能的影响。结果表明,纳米粒子的表面能随尺寸的减小而下降,而对于纳米孔,尺寸的减小将使表面能增加。同时发现,对于纳米粒子,$\gamma(D)/\gamma(\infty)$ 的下降依赖于表面断键数 Z_{b-s},Z_{b-s} 的减小将使 $\gamma(D)/\gamma(\infty)$ 的下降变缓。理论预测与 FCC 晶体结构 Au、Ag、Al、Cu、Ni,HCP 结构 Be、Mg,以及 BCC 晶体结构 Na 的纳米孔及纳米粒子的实验数据和其他理论预测一致。此外,该模型也可用来描述半导体和二元合金的表面能。

第7章 尺寸和形状因素对 Fe_3O_4 纳米晶体 Verwey 转变温度的影响

7.1 Fe_3O_4 纳米晶体的 Verwey 转变背景简介

Verwey 于 1939 年对金属氧化物 Fe_3O_4 进行了研究,发现其在温度约 123 K 的时候,发生了金属—绝缘体转变(也称作 Verwey 转变),123 K 即为 Verwey 转变温度 $T_v(\infty)$,∞ 代表块体材料。从此,Verwey 转变成为广受关注的一个研究领域。M. S. Senn(2012) 研究发现,当温度低于 $T_v(\infty)$ 时,Fe_3O_4 发生了由立方对称结构转变为单斜结构的相变,八面体位置的 Fe^{2+} 和 Fe^{3+} 离子形成三个基态 Fe。相邻的两个单斜晶胞通过晶格扭曲变形形成铁弹性微孪晶偶。这个相变过程伴随着明显的材料性质的突然变化,如电阻率、热容、磁性、延伸率、矫顽力等。Fe_3O_4 在温度为 123 K 时表现出的这些奇特性质吸引了大量的兴趣,在非易失性电阻开关、触发电压、Fe_3O_4 基隧道结、固态能源转换器件等领域都有潜在的应用前景。

近年来,随着纳米科学和纳米技术的发展,对 Fe_3O_4 的 Verwey 转变的研究也拓展到纳米尺度范围。Goya 等人发现,当 Fe_3O_4 纳米粒子的尺寸减小到 50 nm 以下时,其零场冷却曲线中的 Verwey 转变峰消失。此外,Salazar 等人发现,当纳米晶体尺寸减小到低于 22 nm 的时候,由于纳米粒子表面出现 $\gamma-Fe_2O_3$,因此并不能观察到 Verwey 转变。然而,当纳米粒子尺寸 $D=5.5$ nm,其表面吸附脂肪酸时,P. Poddar(2002)发现在温度约为 96 K 的时候出现了 Verwey 转变。通过热容测量法发现,当纳米粒子尺寸为 13 nm 时,

$T_v(D)$ 约为 95 K。2015 年，尺寸范围在 5～100 nm 等计量的 Fe_3O_4 纳米粒子被合成出来，J. Lee 测量发现直到尺寸小于 20 nm 时，Verwey 转变才会出现明显的尺寸依赖效应，并且在尺寸小于 6 nm 时，转变完全消失。Hevroni 通过测量变温扫描隧道谱发现，当纳米粒子尺寸约为 10 nm 时，$T_v(D)$ 约为 101 K。这些报道都表明，对于纳米尺度的 Fe_3O_4，其 Verwey 转变温度 $T_v(D)$ 是尺寸的函数。

此外，纳米粒子的形状因素对 $T_v(D)$ 的影响也受到关注。Mitra 等人制备了 Fe_3O_4 纳米棒，发现由形状诱导的高表面各向异性直接影响其磁共振成像。然而，他们制备的尺寸为 6～14 nm 范围内由{111}面组成高对称八面体形状纳米粒子 $T_v(D)$ 却并没有表现出明显的尺寸依赖性。而 4～13 nm 尺寸范围内的球形纳米粒子并没有观察到 Verwey 转变的发生。

为了解释这些彼此矛盾的实验结果，理解纳米粒子的尺寸效应和形状效应对 Verwey 转变机制的影响，相应的理论工作亟待建立和发展以描述 $T_v(D)$ 函数。通过假设每个分子单元里处在八面体中心点位置(B 点)的两个 Fe 离子贡献 d 轨道的 11 个电子，其中 5 个"自旋向下"最低态吸附 10 个电子，"自旋向上"最低态轨道占有 1 个电子，Cullen 和 Callen(1970)提出，在块体 Fe_3O_4 中带传导是主要的电荷传输机制。在发生相转变的时候，系统的对称性被电荷自身库仑力诱发的有序化所破坏。该理论对宏观 Verwey 转变提出了一个可行的解释，但是并不能对 Verwey 转变给出定量描述。此外，也鲜有理论工作建立 $T_v(D)$ 函数。

7.2 Fe_3O_4 纳米晶体 Verwey 转变温度的尺寸与形状函数模型的建立

有人提出在 $T_v(\infty)$ 以下也可能存在电荷有序。大量的研究发现，Fe_3O_4 在 $T_v(\infty)$ 处发生的有序—无序转变主要是由 $T_v(\infty)$ 以下的库仑长程作用力和在 $T_v(\infty)$ 以上的短程电荷—轨道相关联共同作用所主导的。对于纳米结构，Fe_3O_4 粒子在外壳层将被无序原子所包围，并且尺寸越小，这种无序原子壳越厚。结果，在温度低于$T_v(\infty)$时也可能会观察到电荷有序化，而这和"trimerons"模型的八面体中心点位置(B 点)有关。Verwey 转变温度 $T_v(\infty)$ 可由下式表达：

$$T_v(\infty) = \frac{W(\infty)}{8k_B} \tag{7.1}$$

式中，$W(\infty)$ 定义为将电子从一个晶格阵点移动到另一阵点所需要的能量，实验及理论预测其块体值为 0.1 eV, 0.14 eV 及 0.2 eV; k_B 为玻尔兹曼常量。温度对原子平均热振动能量的影响可表达为 $m(2\pi v_E)^2 \sigma^2(T) = k_B T$，其中 m, v_E, σ 分别为原子质量、爱因斯坦频率和原子热振动的均方根位移。结合公式 (7.1) 可知，在 $T_v(\infty)$ 时，由于最近邻原子间交互作用 $\sigma^2[T_v(\infty)] = k_B T_v(\infty)/[m(2\pi v_E)^2] = W(\infty)/[32m(\pi v_E)^2]$，原子的热振动将破坏电荷—轨道有序化。与此同时，与 Fe(B 点)和 O 离子电荷扰动相耦合的晶格畸变导致的铁弹性微孪晶对称性将被破坏。尽管在立方相中残余晶格畸变会导致在 $T_v(\infty)$ 以上会出现一些短程有序初始结构，但随着立方相的形成整体将会出现电荷无序化。类似在 Lindemann 熔化理论中的假设：当原子的 σ 在 $T_m(\infty)$ 下达到原子直径的分数 f 时，存在 $\sigma^2[T_m(\infty)] = k_B T_m(\infty)/[m(2\pi v_E)^2] = (fh)^2$ 且 $\Theta_D = f[T_m/(mh^2)]^{1/2}$，其中 h 和 Θ_D 分别为原子直径和德拜温度。据此，假设 $W(\infty)/[32(\pi v_E f^2)^2] = K$ (K 为材料常量)，则有 $T_v(\infty) \propto \Theta_D^2(\infty)$。已知 $\Theta_D^2(\infty) \propto E_c(\infty)$，因此 $T_v(\infty) \propto E_c(\infty)$。拓展该关系到纳米尺度，得到

$$T_v(D)/T_v(\infty) = E_c(D)/E_c(\infty) \tag{7.2}$$

结合能的尺寸依赖函数可表示为

$$\frac{E_c(D)}{E_c(\infty)} = \left(1 - \frac{1}{12D/D_0 - 1}\right) \exp\left[\frac{2S_b}{3R(12D/D_0 - 1)}\right] \tag{7.3}$$

式中，R 为理想气体常数，S_b 为块体蒸发熵。D_0 表示 $T_v(D)$ 可能存在的理论最小值，由于 $T_v(D)$ 对纳米晶体表面缺陷非常敏感，实际的临界尺寸将大于 D_0。对于纳米粒子，由于 $4\pi(D_0/2)^2 h =$

$4\pi(D_0/2)^3/3$,因此 $D_0 = 6h = 1.332$ nm。

形状因子 λ 也被引入公式中以描述纳米晶体的形状的影响，因此，Verwey 转变的尺寸和形状依赖函数 $T_v(\lambda, D)$ 可由下式给出：

$$\frac{T_v(\lambda, D)}{T_v(\infty)} = \left(1 - \frac{1}{12D/D_0 - 1}\right) \exp\left[-\frac{2\lambda S_b}{3R(12D/D_0 - 1)}\right] \quad (7.4)$$

7.3 Fe_3O_4 纳米晶体 Verwey 转变温度理论模型与实验结果的对比与分析

图 7.1 给出了公式(7.4)对形状因子 $\lambda = 1$ 和维度 $d = 0$ 的球形纳米粒子的预测以及实验数据对比。由于 $S_b(\infty)$ 值缺少相关数据，因此近似取 $S_b(\infty) \approx 13R$。图中 ★、×、●、◇、□、△、◆ 为对应的实验值。$T_v(D)$ 在大于 20 nm 时并没有明显的尺寸依赖性，但是当 $D < 20$ nm 时，随着尺寸的减小，$T_v(D)$ 急剧下降。当纳米粒子尺寸分别为 5.5 nm、10 nm 及 13 nm 时，理论计算得到的 $T_v(D)$ 值分别为 93.8 K、108 K 和 111.6 K，这与对应的实验结果 96 K、101 K 及 95 K 相吻合。由于 $T_v(D)$ 对 O 元素含量计量比极度敏感，此处引用的都是 Fe_3O_4 纳米晶体等计量比的实验数据，同时图 7.1 也只考虑了尺寸 D。该模型对实验结果的精确描述也验证了在 $T_v(\infty)$ 下的电荷—轨道的有序—无序转变可以直接与原子的热振动相关联，并且这种关联在纳米尺度也同样有效。众所周知，随着尺寸 D 的减小，表面体积比 δ 急剧增加，纳米晶体表面出现大量的失配原子和缺陷，从而破坏了表面结构的对称性。所有这些特征最终导致纳米晶体表面原子不稳定性增加和热振动增强。结果，表面原子结构对称性发生改变，在 $T_v(D)$ 下原子的电荷—轨道也更容易发生由长程有序向短程有序—无序的转变。由于随着 D 的减小，$T_v(D)$ 急剧下降，这种有序—无序转变对小尺寸纳米晶体尤为显著。因此，可以预见，当 D 足够小的时候，纳米晶体表面完全被无序原子或非等计量比的缺陷结构占据，电荷—轨道的长程

有序消失,从而在实验上更难观察到 $T_v(D)$。同时需要注意,纳米晶体表面原子结构的无序化同时也深刻影响着磁矩,这对理解 Fe_3O_4 纳米晶体的磁性和电子传输性质具有重要意义。由公式(7.3)知,$E_c(D)$ 随着 D 的减小而下降,由原子热振动主导的无序化驱动力或 σ 增加,而由磁矩交互作用主导的电荷有序化驱动力下降。结果使纳米晶体表面处单位质量拥有的磁矩比纳米晶体核心区域少得多,这种磁矩钉扎效应对巨磁电阻和高场巨磁电阻弱饱和现象都有重要影响。

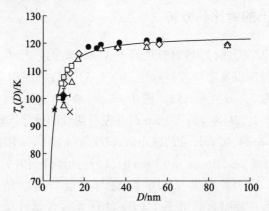

图 7.1 预测的 $T_v(D)$ 函数与对应纳米粒子实验值对比

图 7.2 为参考 Mitra 的实验结果考虑形状效应对 $T_v(D)$ 的影响。图中,●,★为文献[308]和[303]中测量的实验值,而○与☆由公式(7.4)计算得到,其中 $\lambda_{八面体} = 1.23$,$\lambda_{球体} = 1$。Mitra 发现,对于尺寸为 6 nm,8 nm 和 10 nm 的八面体纳米粒子,测量得到的 $T_v(D)$ 值约为 125 K,即与块体值相当。而由于存在非等计量比的表面氧化层,尺寸在 4~13 nm 范围的球状纳米粒子并没有观察到 $T_v(D)$。很明显,这与前面提到的文献结果相矛盾。在图 7.2 中,符号●为测量 6 nm 八面体形状纳米粒子的 $T_v(D)$ 值,符号★为引用的 5.5 nm 的球状纳米粒子对应的实验值。对于 5.5 nm 球形纳米粒子,$T_v(D)$ 的预测值为 93.8 K,这与其实验值 96 K 非常接近。而对于 6 nm 的八面体形状的纳米粒子,$T_v(D)$ 计算值为 91.4 K,远

比块体值 120 K 低。实验上测量得到的较高 $T_v(D)$ 值可能是由在制备 Fe_3O_4 纳米晶体时在纳米晶体表面残存密堆积的氩分子层所造成的。此外,通过图 7.2 我们也可以看出,当尺寸小于 20 nm 时,八面体形状纳米粒子的 $T_v(D)$ 的下降要比对应球状纳米粒子值快得多。这很容易理解,因为形状因素直接确定了表面原子数,表面原子配位数的减小将直接影响纳米晶体的结合能 $E_c(D)$。根据公式(7.4),$T_v(D,\lambda)$ 随着 λ 的增加和 D 的减小而下降。由于 $\lambda_{八面体} > \lambda_{球状}$,八面体形状纳米粒子拥有更大的表面体积比 δ、更多的断键数及更低的结合能 $E_c(D)$,根据公式(7.2),$T_v(D)_{球状} > T_v(D)_{八面体}$。

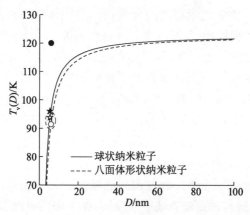

图 7.2 八面体形状和球状纳米粒子 $T_v(D)$ 值对比

7.4 本章小结

综上,本章提出了无任何可调参数的解析模型描述尺寸和形状因素对 Fe_3O_4 纳米晶体的 Verwey 转变温度 $T_v(D,\lambda)$ 的影响。结果表明,$T_v(D,\lambda)$ 随着形状因子 λ 的增加及尺寸 D 的减小而下降。理论模型的预测与实验结果相吻合,这也意味着 $T_v(D)$ 可在块体和纳米尺度整个范围内与原子热振动相关联。本章中的计算和预测将为未来能源转换和内存器件的设计提供理论指导。

第 8 章 尺寸、维度和压力对 CdSe 和 ZnO 纳米晶体能带的调制效应

8.1 纳米晶体能带简介

近年来,由于纳米材料所展现出的与块体材料截然不同的物理、化学和机械性质,对其的研究日益受到关注。当材料尺寸 D 减小到纳米尺度时,表面体积比(A/V)急剧增加,表面出现大量配位缺陷,最终导致其电子结构和光学性质的改变。对于半导体纳米晶体,包括 IV (Si)族、Ⅲ-V (GaN, GaP, GaAs, InP, InN 等)族,以及 Ⅱ-VI (ZnS, ZnSe, CdS, CdSe 等)族闪锌矿(ZB)和纤锌矿(WZ)结构半导体,直接的结果就是能带展宽 $\Delta E_g = E_g(D) - E_g(\infty)$,其中 ∞ 表示块体尺寸,这对改变其电子结构和光学性质起到重要的作用。同时,通过调制能隙改变纳米晶体电子结构可以拓展纳米材料在更多领域中的应用,如太阳能电池、高速场效应晶体管、激光二极管等。

研究发现,除了尺寸效应,纳米晶体的维度同样也影响 E_g。例如,纳米线和纳米薄膜的维度分别为 1 和 2,而纳米粒子的维度为 0,由于尺寸效应 D 和维度效应 d 都对表面体积比 A/V 做贡献,而表面键收缩可直接改变表面原子的结合能,因此,纳米晶体的能隙可以用函数 $E_g(D,d)$ 来描述。此外,外压 P 同样会改变原子间的结合能,所以能隙的函数最终可表示为 $E_g(D,d,P)$ 形式。压力对能隙 E_g 的影响可由经验表达式 $E_g(P) = E_g(0) + AP^2 + BP$ 给出,其中 A 和 B 为可调参数,但其物理意义并不清楚。最近,基于 BOLS 理论及局域键平均方法(LBA),通过考虑压力 P 对配位原子

的键能和键长 Hamiltonian 扰动的反应,孙长庆等人(2009,2010)提出了无可调参数的理论模型描述 $\Delta E_g(D,d,P)$ 函数,其预测结果与实验数据相吻合。然而,维度 d 对 $E_g(D)$ 和 $E_g(D,P)$ 函数影响的物理起源仍旧不清楚,尤其是尺寸 D、维度 d 和压力 P 对 E_g 影响的竞争性关系需要进一步阐明。因此,本章将提出新模型描述 $\Delta E_g(D,d,P)$ 函数,发现由不同表面体积比 A/V 产生的键能差异是影响 $E_g(D,P)$ 和 $E_g(D,d,P)$ 函数的本质原因。对 CdSe 和 ZnO 纳米晶体能隙数据的预测验证新模型的准确性。

8.2 纳米晶体能带的尺寸、维度和压力函数模型的建立

基于近自由电子方法,能隙 E_g 和原子结合能相关函数已经发展出来,其可以成功描述Ⅲ-Ⅴ和Ⅱ-Ⅵ族纳米半导体材料的能隙值,表达式为

$$\frac{E_g(D,d)}{E_g(\infty)} = 2 - \frac{E_c(D,d)}{E_c(\infty)} \tag{8.1}$$

而尺寸 D 和维度 d 依赖的原子结合能可表达为

$$\frac{E_c(D,d)}{E_c(\infty)} = \left[1 - \frac{1}{(12D/D_0)-1}\right]\exp\left[-\frac{2S_b}{3R}\frac{1}{(12D/D_0)-1}\right] \tag{8.2}$$

维度 d 的影响可由 D_0 描述,即

$$D_0 = 2(3-d)h \tag{8.3}$$

式中,$d=0,1,2$ 分别表示纳米粒子、纳米线和纳米薄膜。

由于纳米晶体结构是不依赖于尺寸的,假设压力 P 对纳米晶体的结合能 E_c 的影响与块体相同,则有

$$\frac{E_c(D,d,P)}{E_c(\infty,P)} = \frac{E_c(D,d,0)}{E_c(\infty,0)} \tag{8.4}$$

压力对结合能的影响 $E_c(\infty,P)$ 可表示为

$$E_c(\infty,P) = E_c(\infty,0) + \Delta E_c(\infty,P) \tag{8.5}$$

其中 $\Delta E_c(\infty,P) = -\int_{V_0}^{V} P(V)\mathrm{d}V$,$P(V)$ 可根据 Birch Murnaghan 绝

热状态方程得出：

$$P(V) = \frac{3B_0}{2}\left[\left(\frac{V}{V_0}\right)^{-\frac{7}{3}} - \left(\frac{V}{V_0}\right)^{-\frac{5}{3}}\right]\left\{1 + \frac{3}{4}(B_0' - 4)\left[\left(\frac{V}{V_0}\right)^{-\frac{2}{3}} - 1\right]\right\} \tag{8.6}$$

由公式(8.1),(8.3),(8.4)和(8.5)可得

$$\frac{E_c(D,d,P)}{E_c(\infty,0)} = \left[1 - \frac{1}{(12D/D_0) - 1}\right]\exp\left[-\frac{2S_b}{3R}\frac{1}{(12D/D_0) - 1}\right]$$

$$\frac{[E_c(\infty,0) + \Delta E_c(\infty,P)]}{E_c(\infty,0)} \tag{8.7}$$

因此，

$$\frac{E_g(D,d,P)}{E_g(\infty,0)} = 2 - \frac{E_c(D,d,P)}{E_c(\infty,0)} \tag{8.8}$$

8.3 纳米晶体能带函数理论模型和 CdSe，ZnO 的实验结果的对比与分析

图 8.1 所示为维度 $d = 0$，尺寸 D 分别为 2.4 nm 和 3.5 nm 的 CdSe 纳米粒子的 $E_g(D,d,P)$ 函数曲线。图中，●和▲对应实验值。计算过程中用到的参数有：$V_0 = 112.2$ Å^3，$B_0 = 53.3$ GPa，$B_0' = 4$ GPa，$E_c = 2.387$ eV/atom，$S_b = 67.63$ J/(g - atom · K)，$h = 0.219$ nm，$E_g(\infty) = 1.74$ eV。其中，B_0 为体积模量，B_0' 为其一阶导数。很明显 $E_g(2.4 \text{ nm}) > E_g(3.5 \text{ nm})$，这是由尺寸效应引起的。随着压力 P 的增加，体积 V 逐渐减小，能隙也逐渐展宽，这说明压力 P 的增加与尺寸 D 的减小对 $E_g(D,P)$ 做了同等的贡献，这与其他文献报道结果一致。通过分析压力 P 和尺寸 D 对键能变化的影响可以解释这个趋势。当尺寸减小时，纳米晶体表面出现大量配位缺陷和断键，使表层原子键长收缩，键能增强，这也是使能隙展宽的本质原因所在。根据公式(8.7)，V 在 P 的作用下被压缩，这也导致表层原子的键长和键强出现与 D 减小时一样的变化趋势。此外，体积 V 的压缩也减小了纳米晶体的尺寸 D，这同样也诱导能隙

E_g 的展宽。

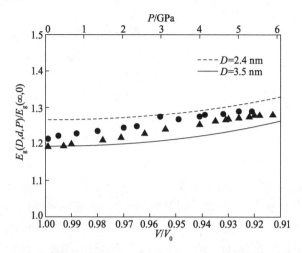

图 8.1 尺寸分别为 2.4 nm 和 3.5 nm 的 CdSe 纳米粒子的 $E_g(D,d,P)$ 函数曲线

图 8.2 为压力对不同尺寸和维度的 ZnO 纳米晶体的能隙的影响曲线。图中，△，◆，●及▲代表实验值。计算过程中用到的参数有：$V_0 = 48$ Å^3，$B_0 = 160$ GPa，$B_0' = 4.4$，$E_c = 2.829$ eV/atom，$S_b = 13R$，$h = 0.166$ nm 及 $E_g(\infty) = 3.26$ eV。图 8.2 中，压力 P 同时展宽了 ZnO 纳米粒子和纳米线的 $E_g(D,d,P)$，这与图 8.1 的 CdSe 结果一致。同时注意到，对于纳米粒子和纳米线，它们的尺寸 D 存在非常大的跨度 (30 ~ 100 nm)，然而，它们的能隙值差 $E_g(D,d,0)$ 变化却非常小，即 $\Delta E_g(D,d,0)/E_g(\infty,0) = E_g(30 \text{ nm},0,0)/E_g(\infty,0) - E_g(100 \text{ nm},1,0)/E_g(\infty,0) = 0.018$，尺寸和维度效应对能隙的影响并不明显。

图 8.3 中，△，◇，□，◁，+，☆，○及×引自文献[321 – 328]，描述 $P = 0$ GPa 时纳米粒子的变化趋势；▽引自文献[328]，描述 $P = 10.66$ GPa 时纳米粒子的变化趋势；▲引自文献[329]，描述 $P = 0$ GPa 时 ZnO 纳米线的变化趋势，⊕引自文献[330]，描述 ZnO 纳米薄膜的变化趋势。通过分析图 8.3 我们发现，纳米粒子和纳

米线的能隙值 $E_g(D,d,P)$ 只有在 $D < 10$ nm 才会明显,尤其当尺寸小于 5 nm 时,能隙值急剧增加,它们之间的差别也立显。维度效应直接与 A/V 相关联,对于纳米粒子、纳米线和纳米薄膜,分别有 $A/V = 6/D$,$A/V = 4/D$ 及 $A/V = 2/D$。尺寸改变 ΔD 将引起表面体积比 A/V 变化为:纳米粒子,$6/\Delta D$;纳米线,$4/\Delta D$;纳米薄膜,$2/\Delta D$。也就是说,A/V 对尺寸 D 的敏感性遵循:纳米薄膜<纳米线<纳米粒子。当纳米粒子尺寸减小到确定值 D^0 时,对应能隙值为 $E_g^0(D^0,0,0)$,而对于纳米线和纳米薄膜,能隙值达到 $E_g^0(D^0,0,0)$ 将需要牺牲更大的尺寸 $D < D^0$ 以补偿 A/V 的不足。这在图 8.3 中 $P = 0$ GPa 的情况体现得非常明显。对于尺寸为 5.1 nm 的纳米粒子,$E_g(D,0,0)/E_g(\infty,0) = 1.147$,而达到该能隙值时的纳米线直径为 3.4 nm,纳米薄膜厚度为 1.7 nm。图 8.3 同时给出了 $P = 10.66$ GPa 时的情况。显然,随着尺寸 D 减小,增加 P 同样会提高 $E_g(D,d,P)$ 并削弱尺寸的贡献,因此函数曲线变得更加平缓。但是当 $D < 2.5$ nm 时,P 的贡献被削弱了,$E_g(D,d)$ 的增加几乎全部由尺寸所贡献。这可由键能对尺寸的依赖性来解释。压力 P 对纳米晶体提供的附加能只要体现在施加于每个键上,随着尺寸的减小,A/V 急剧增加,键主要集中分布在表面原子层。大量的断键存在大大增强了表层原子的键能,从而使小尺寸的纳米晶体更加难以压缩,因此,压力 P 对键能变化的贡献也就减小。从图 8.3 中还可以发现,P 诱导的能隙增加 $\Delta E_g(D,d,P)$ 遵循:纳米薄膜 > 纳米线 > 纳米粒子。例如,当尺寸 $D = 5$ nm 时,计算 $\Delta E_g = E_g(D,d,P)/E_g(\infty,0) - E_g(D,d,0)/E_g(\infty,0)$ 的值,纳米薄膜为 0.0530,纳米线为 0.0502,纳米粒子为 0.0449。这种关系可由维度 d 和压力 P 对键能的影响来解释。对于纳米粒子、纳米线及纳米薄膜,A/V 可分别表示为 $6/D$,$4/D$ 和 $2/D$,其中纳米粒子拥有最大的表面曲率、最多的表层断键数、最大的键收缩和最强的键能,而纳米薄膜则相反。因此对于 P 的敏感度,纳米粒子最低,纳米薄膜最高。即压力对不同维度的纳米晶体提供的附加能遵循:纳米粒子<纳米线<纳米薄膜。根据公式(8.7)和(8.8),纳米粒子 $D_0 =$

$6h$,纳米线 $D_0=4h$,纳米薄膜 $D_0=2h$,因此 d 和 P 对 $E_g(D,d,P)$ 的影响就可以理解了。

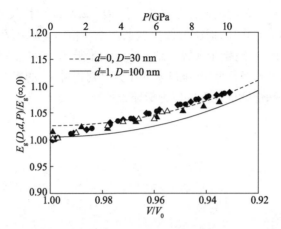

图 8.2　压力对不同尺寸和维度的 ZnO 纳米晶体的 $E_g(D,d,P)$ 函数的影响

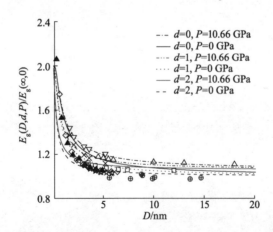

图 8.3　不同维度不同压力 P 下 ZnO 的 $E_g(D,d,P)$ 函数曲线与实验结果对比

8.4　本章小结

本章主要讨论压力对不同维度纳米晶体的能隙的影响。发现

尺寸 D、维度 d（由 A/V 表达）及 P 作用导致的纳米晶体表层原子的键能变化是改变 E_g 的决定因素。表面体积比 A/V 越大，$E_g(D,d,P)$ 对 P 的依赖性就越强。当 $D<2.5$ nm 时，尺寸效应对能隙值起主导作用，而压力效应变弱。能隙值 $E_g(D,d,P)$ 对压力 P 的敏感性遵循：纳米薄膜 > 纳米线 > 纳米粒子。这些结果有利于我们更好地理解尺寸 D、维度 d 和压力 P 对能隙 E_g 的影响，为量子器件设计和性能优化提供理论指导。

第 9 章 总结与展望

通过热力学方法，本书主要研究了在介观尺度领域中双金属合金的有序—无序转变、熔化转变、液晶材料的熔化相变、氧化物 In_2O_3 的固-固相变，以及相变压力的尺寸效应。通过讨论，我们得到了以下结论：

（1）通过引入形状参数，发现形状对双金属合金的有序—无序转变及熔化转变主要是受其表面/体积比变化的影响，正是由于纳米晶体表面体积比的急剧增加才导致相变温度减小。对于有序—无序转变，我们发现不同维度和形状的转变温度遵循 T_{OD}（薄膜）$> T_{OD}$（片状颗粒）$> T_{OD}$（线）的顺序，尺寸对相变的影响要远大于形状。对于 Au-Pt 的熔化转变，随着 Pt 元素的增加合金熔化温度提高，这主要是由于 Pt 为高熔点元素，Pt-Pt 之间的结合能要大于 Au-Au，导致纳米颗粒变得更加稳定。

（2）通过所做的工作，成功把纳米晶体熔化热力学模型的研究对象拓展到液晶材料领域，描述了 PAA 和 5CB 液晶材料相变的尺寸效应，其晶体—液晶体、液晶体—液体、液体—液晶体和液晶体—晶体的相变温度都随着束缚液晶体的纳米微孔的尺寸减小而降低，而且当纳米微孔被惰性分子修饰之后，其相变温度都降低，这主要是由液晶与纳米微孔界面间的氢键密度所决定的。惰性分子的加入不但破坏了界面间的大部分氢键，而且引入了缺陷，这都导致界面处的液晶分子更容易移动而发生有序—无序转变，从而降低了相应的相变温度。

（3）研究 In_2O_3 纳米晶体的相变温度和相变压力的尺寸效应发现，相变温度是随着尺寸的下降而减小的，这与实验观察的结果一致；但是也同样发现了相变压力随尺寸下降而减小的趋势，与实验观察到相变压力不发生变化的结果截然不同。这可能是由于低温状态下的一些动力学因素导致的相变的迟缓。分析了表面能与表面应力在相变过程中所起的作用发现，随着尺寸的下降，表面能是趋向减小相变温度和相变压力的，为相变驱动力的主导因素；而表面应力则趋向增强相变温度和相变压力，为相变驱动力的次要因素。详细分析讨论了它们对纳米相变体系的总能量变化的贡献。

（4）通过观察一些实验结果，发现纳米晶体的相变压力随着尺寸的减小有着不同的变化趋势，也就是 $P_e(D)$ 可能大于或者小于 $P_e(\infty)$。研究发现，$P_e(D)$ 的变化趋势可以通过对比两相的表面能 γ 表面应力 f 和原子单位体积 V_g 来预测。而且 $P_e(D)$ 的变化趋势，对 D 依赖的强度，以及 γ 和 f 对 $P_e(D)$ 单独贡献的角色，都可以通过 $3\Delta(\gamma V) + 2\Delta(fV)$ 的值及 $-3\Delta(\gamma V)$ 和 $-2\Delta(fV)$ 相对值的大小来评判。

（5）通过建模描述了纳米晶体表面能的变化趋势。随着纳米晶体尺寸的减小，纳米粒子的表面能是下降的，而对于纳米孔洞，其表面能是增加的。这与当尺寸减小时纳米粒子表面呈现的键收缩和纳米孔洞表面呈现的键拉伸相关。通过比对理论模型和 FCC、BCC、HCP 纳米粒子的实验数据，Cu 纳米粒子与纳米孔洞的计算机模拟数据，Si 纳米孔及 Cu-Al、Cu-Au 合金的理论计算结果，得到了很好的一致性。此外，通过分析直径为 3.8 nm 的纳米晶体 Au 的 Au(111)、Au(100) 及 Au(110) 表面能，发现表面能随着表面断键数 Z_{b-s} 的减小而增加，即对于这三种 Au 表面，由于 $[Z_{b-s}^{(111)} = 3] < [Z_{b-s}^{(100)} = 4] < [Z_{b-s}^{(110)} = 5]$，则表面能遵循 $\gamma(111) > \gamma(100) > \gamma(110)$，这与实验结果相一致。

（6）Verwey 转变是 Fe_3O_4 晶体的一个重要物理性质，实验发现当其尺寸减小到纳米尺度的时候，Verwey 转变温度也呈现出尺寸依赖性质并随尺寸减小而下降。通过引入结合能尺寸效应函数，建立了描述 Verwey 转变温度的尺寸效应函数，并讨论了形状因素的影响。由于外界因素影响,虽然在实验上并未观察到形状因素对 Verwey 转变温度的影响,但是建立的理论模型预测显示八面体形状的 Fe_3O_4 纳米粒子拥有比球状粒子更低的 Verwey 转变温度,而这与由表面体积比所决定的形状因子 λ 有关。λ 越大,Verwey 转变温度越低。

(7) 通过建模,考查了 CdSe 和 ZnO 纳米晶体能隙值受纳米晶体尺寸、维度和压力的影响。结果表明,由尺寸 D、维度 d 和压力 P 诱导的纳米晶体表层原子键能的变化是导致能隙 E_g 发生变化的本质原因。并且表面体积比 A/V 值越大,$E_g(D,d,P)$ 对压力 P 的依赖性越强。因此,不同维度的纳米晶体 $E_g(D,d,P)$ 对 P 的敏感性遵循关系:纳米薄膜 > 纳米线 > 纳米粒子。

目前,随着纳米科学和纳米技术研究的不断深入,纳米材料热力学理论也日趋成熟。本书仅针对形状、维度、压力、界面键合、表面配位等几种因素对材料有序—无序转变、熔化转变、液晶相变、固态相变、表面能、Verwey 转变温度、能隙值等材料性质预测值修正而进行的浅显初探。在应用过程中,量子器件所处的环境、制备方法、服役时间等都会对器件的性质产生重要影响,从而使纳米材料热力学理论的建立变得复杂,为了使得热力学理论精确描述纳米材料性质的变化趋势,应变、缺陷、界面条件、外界温度等因素都应加以考虑以对经典理论模型进行修正,因此,还有大量的后续工作需要完成。此外,伴随着石墨烯材料研究的发展,大量 2D 材料相继涌现,将纳米材料热力学理论拓展到描述 2D 材料的各种性质行为变化也是一项极具挑战性的工作,这也是作者未来科研工作的方向。

参考文献

[1] TANIGUCHI N. On the basic concept of nanotechnology [C]. Proceeding of international conference on production engineering. Tokyo: JSPE, 1974, 18-23.

[2] HILL T L. Thermodynamics of small systems [M]. Part II. New York: Dover publications, Inc. 1994, 167.

[3] FEYNMANN R P. There's plenty of room at the bottom [J]. Engineering and Science, 1960, 23: 22-36.

[4] GLEITER H, HANSEN N, HORSEWELL A, et al. In Proceedings of the second Risφ international symposium on metallurgy and materials science. Roskilde: Risφ National Laboratory, 1981, 15.

[5] MYERS D. Surfaces, interfaces, and colloids: Principles and applications [M]. 2nd ed. New York: John Wiley & Sons, Inc. , 1999.

[6] GLEITER H. Nanostructured materials: Basic concept and microstructure [J]. Acta Materialia, 2000, 48: 1.

[7] CAVICCHI R E, SILSBEE R H. Coulomb suppression of tunneling rate from small metal particles [J]. Physical Review Letters, 1984, 52: 1453.

[8] TAKAGI M. Electron-diffraction study of liquid-solid transition of thin metal films [J]. Journal of the Physical Society of Japan, 1954, 9: 359.

[9] Ball P, Garwin L. Science at the atomic scale [J]. Nature,

1992, 355: 761.

[10] KASHCHIEV D. Thermodynamically consistent description of the work to form a nucleus of any size [J]. Journal of Chemical Physics, 2003, 118: 1837.

[11] WELLER D, MOSER A, FOLKS L, et al. High K_u materials approach to 100 Gbits/in^2 [J]. IEEE Transactions on Magnetics, 2000, 36: 10.

[12] PLUMER M L, VAN EK J, WELLER D. The physics of ultra-high-density magnetic recording [M]. Berlin: Springer, 2001.

[13] SELLMYER D J, YU M, KIRBY R D. Nanostructured magnetic films for extremely high density recording [J]. Nanostructure Materials, 1999, 12: 1021 – 1026.

[14] YU M, LIU Y, SELLMYER D J. Nanostructure and magnetic properties of composite CoPt: C films for extremely high-density recording [J]. Journal of Applied Physics, 2000, 87: 6959 – 6961.

[15] WELLER D, MOSER A. Thermal effect limits in ultrahigh-density magnetic recording [J]. IEEE Transactions on Magnetics, 1999, 35: 4423 – 4439.

[16] SAKUMA A. First principle calculation of the magnetocrystalline anisotropy energy of FePt and CoPt ordered alloys [J]. Journal of Physical Society of Japan, 1994, 63: 3053 – 3058.

[17] HIMPSEL F J, ORTEGA J E, MANKEY G J, et al. Magnetic nanostructures [J]. Journal of Advanced Physics, 1998, 47: 511 – 597.

[18] SUN S H, MURRAY C B, WELLER D, et al. Monodisperse FePt nanoparticles and ferromagnetic FePt nanocrystals superlattices [J]. Science, 2000, 287: 1989.

[19] WATANABE M, MASUMOTO T, PING D H, et al. Microstructure and magnetic properties of FePt-Al-O granular thin films [J].

Applied Physics Letters, 2000, 76: 3971.

[20] BIAN B, LAUGHLIN D E, SATO K, et al. Fabrication and nanostructure of oriented FePt particles [J]. Journal of Applied Physics, 2000, 87: 6962.

[21] YAN M L, ZENG H, POWERS N, et al. $L1_0$, (001) -oriented FePt: B_2O_3 composite films for perpendicular recording [J]. Journal of Applied Physics, 2002, 91: 8471.

[22] KUO P C, CHEN S C, YAO Y D, et al. Microstructure and magnetic properties of nanocomposite FePtCr-SiN thin films [J]. Journal of Applied Physics, 2002, 91: 8638.

[23] SAITO T, KITAKAMI O, SHIMADA Y. Thickness dependence of the phase transformation in FePt alloy thin films [J]. Journal of Magnetism Magnetic Materials, 2002, 239: 310 – 313.

[24] TSKAHASHI Y K, OHKUBO T, OHNUMA M, et al. Size effect on the ordering of FePt granular films [J]. Journal of Applied Physics, 2003, 93: 7166.

[25] TSKAHASHI Y K, KOYAMA T, OHNUMA M, et al. Size dependence of ordering in FePt nanoparticles [J]. Journal of Applied Physics, 2004, 95: 2690.

[26] SUN C Q. Size dependence of nanostructures: Impact of bond order deficiency [J]. Progress in Solid State Chemistry, 2007, 35: 1 – 159.

[27] SUN C Q, SHI Y, LI C M, et al. Size-induced undercooling and overheating in phase transitions in bare and embedded clusters [J]. Physical Review B, 2006, 73: 075408.

[28] ZHONG W H, SUN C Q, LI S, et al. Impact of bond order loss on surface and nanosolid magnetism [J]. Acta Materialia, 2005, 53: 3207 – 3214.

[29] JIANG R, ZHOU Z F, YANG X X, et al. Size-depressed critical temperatures for the order-disorder transition of FePt, CoPt,

FePb, Cu_2S, and ZnS nanostructures [J]. Chemical Physics Letters, 2013, 555: 202 -205.

[30] OUYANG B, QI W H, LIU C Z, et al. Size and shape dependent order-disorder phase transition of Co-Pt nanowire [J]. Computational Materials Science, 2012, 63: 286 -291.

[31] QI W H, HUANG B Y, WANG M P, et al. Generalized bond-energy model for cohesive energy of small metallic particles [J]. Physics Letters A, 2007, 370: 494.

[32] QI W H, WANG M P, LIU Q H. Shape factor of nonspherical nanoparticles [J]. Journal of Materials Science, 2005, 40: 2737 -2739.

[33] LI Y J, QI W H, HUANG B Y, et al. Generalized Bragg-Williams model for the size-dependent order-disorder transition of bimetallic nanoparticles [J]. Journal of Physics D: Applied Physics, 2011, 44: 115405.

[34] XIONG S Y, QI W H, HUANG B Y, et al. Size and shape dependent Gibbs free energy and phase stability of titanium and zirconium nanoparticles [J]. Materials Chemistry Physics, 2010, 120: 446 -451.

[35] XIONG S Y, QI W H, HUANG B Y, et al. Gibbs free energy and size-temperature phase diagram of hafnium nanoparticles [J]. Journal of Physical Chemistry C, 2011, 115: 10365 - 10369.

[36] XIONG S Y, QI W H, CHENG Y J, et al. Modeling size effect on the surface free energy of metallic nanoparticles and nanocavities [J]. Physical Chemistry Chemical Physics, 2011, 13: 10648 -10651.

[37] LI Y, QI W H, HUANG B Y, et al. Modeling the size-and shape-dependent surface order-disorder transition of $Fe_{0.5}Pt_{0.5}$ nanoparticles [J]. Journal of Physical Chemistry C, 2012, 116:

26013 - 26018.

[38] QI W H, LI Y J, XIONG S Y, et al. Modeling size and shape effects on the order-disorder phase-transition temperature of CoPt nanoparticles [J]. Small, 2010, 6: 1996 - 1999.

[39] ALLOYEAU D, RICOLLEAU C, MOTTET C, et al. Size and shape effects on the order-disorder phase transition in CoPt nanoparticles [J]. Nature Materials, 2009, 8: 940 - 946.

[40] DASH J G. History of the search for continuous melting [J]. Review of Modern Physics, 1999, 71: 1737.

[41] CHRISTIAN J W. The theory of transformations in metals and alloys [M]. 2nd ed. Part I, Equilibrium and General Kinetic Theory. Oxford: Pergamon Press, 1975: 418 - 475.

[42] GLEITER H. Nanocrystalline materials [J]. Progress in Material Science, 1989, 33: 223.

[43] LU L, SUI M L, LU K. Superplastic extensibility of nanocrystalline copper at room temperature [J]. Science, 2000, 287: 1463.

[44] LINDEMANN F A. Üder die berechnung molekularer eigenfrequenzen [J]. Physik. Zeits, 1910, 11: 609.

[45] BORN M. Thermodynamics of crystals and melting [J]. Journal of Chemical Physics, 1939, 7: 591.

[46] GORECKI T. Vacancies for solid krypton bubble copper, nickel and gold particles [J]. ZEITSCHRIFT FÜR METALLKUNDE, 1974, 65: 426.

[47] RUBCIC A, RUBCIC J B. Simple model for volume change at melting of elements [J]. Fizika, 1978, 10 (Suppl. 2): 294.

[48] GRABACK L, BOHR J. Superheating and supercooling of lead precipitates in aluminum [J]. Physical Review Letters, 1990, 64: 934.

[49] LU K, LI Y. Homogeneous nucleation catastrophe as a kinetic

stability limit for superheated crystals [J]. Physical Review Letters, 1998, 80: 4474.

[50] ROSSOUW C J, DONNELLY S E. Superheating of small solid – argon bubbles in aluminum [J]. Physical Review Letters, 1985, 55: 2960.

[51] LAI S L, GUO J Y, PETROVA V, et al. Size dependent melting properties of small Tin particles: Nanocalorimetric measurements [J]. Physical Review Letters, 1996, 77: 99.

[52] ZHANG D L, CANTOR B. Melting behaviour of In and Pb particles embedded in an Al matrix [J]. Acta Materialia, 1991, 39: 1595.

[53] GOSWAMI R, CHATTOPADHYAY K. The superheating and the crystallography of embedded Pb particles in f. c. c. Al, Cu and Ni matrices [J]. Acta Metallurgica et Materialia, 1995, 43: 2837.

[54] SHENG H W, REN G, PENG L M, et al. Superheating and melting-point depression of Pb nanoparticles embedded in Al matrixes [J]. Philosophical Magazine Letters, 1996, 73: 179.

[55] GOSWAMI R, CHATTOPADHYAY K. The superheating of Pb embedded in a Zn matrix: The role of interface melting [J]. Philosophical Magazine Letters, 1993, 68: 215.

[56] ANDERSEN H H, JOHNSON E. Structure, morphology and melting hysteresis of ion-implanted nanocrystals [J]. Nuclear Instruments and Methods in Physics Research Section B, 1995, 106: 480.

[57] BUFFAT P A, BOREL J P. Size effect on the melting temperature of gold particles [J]. Physical Review A, 1976, 13: 2287.

[58] ALLEN G L, BAYLES R A, GILES W W, et al. Small particle melting of pure metals [J]. Thin Solid Films, 1986, 144: 297.

[59] COOMBES C J. The melting of small particles of lead and indi-

um [J]. Journal of Physics F: Metal Physics, 1972, 2: 441.

[60] SKRIPOV V P, KOVERDA V P, SKOKOV V N. Size effect on melting of small particles [J]. Physica Status Solidi A, 1981, 66: 109.

[61] PEPPIATT J, SAMBLES J R. The melting of small particles (Ⅱ) [M]. Bismuth, Proceedings of the Royal Society of London Series A, 1975, 345: 387.

[62] KOFMAN R, CHEYSSAC P, GARRIGOS R. From the bulk to clusters: Solid-liquid phase transitions and precursor effects [J]. Phase Transitions, 1990, 24: 283.

[63] CASTRO T, REIFENBERGER R, CHOI E, et al. Size-dependent melting temperature of individual nanometer-sized metallic clusters [J]. Physical Review B, 1990, 42: 8548.

[64] BOREL J P. Thermodynamical size effect and the structure of metallic clusters [J]. Surface Science, 1981, 106: 1.

[65] SOLLIARD C. Debye-Waller factor and melting temperature in small gold particles realted size effect [J]. Solid State Communications, 1984, 51: 947.

[66] EVENS J H, MAZEY D J. Evidence for solid krypton bubbles in copper, nickel and gold at 293 K [J]. Journal of Physics F: Metal Physics, 1985, 15: L1.

[67] BECK T L, JELLINEK J, BERRY R S. Rare gas clusters: Solids, liquids, slush, and magic numbers [J]. Journal of Chemical Physics, 1987, 87: 545.

[68] GOLDSTEIN A N, ECHER C M, ALIVISATOS A P. Melting in semiconductor nanocrystals [J]. Science, 1992, 256: 1452.

[69] GOLDSTEIN A N. The melting of silicon nanocrystals: Submicron thin-film structures derived from nanocrystal precursors [J]. Applied Physics A, 1996, 62: 33.

[70] SEMENCHENKO V K. Surface phenomena in metals and

alloys [M]. Oxford: Pergamon, 1961: 281.

[71] SAMBLES J R. The extrusion of liquid between highly elastic solid [J]. Proceedings of the Royal Society A, 1971, 324: 339.

[72] COUCHMAN P R, JESSER W A. Thermodynamic theory of size dependence of melting temperature in metals [J]. Nature, 1977, 269: 481.

[73] ALLEN G L, GILE W W, JESSER W A. The melting temperature of microcrystals embedded in a matrix [J]. Acta Metallurgica, 1980, 28: 1695.

[74] ECKERT J, HOLZER J C, AHN C C, et al. Melting behavior of nanocrystalline aluminum powders [J]. Nanostructured Materials, 1993, 2: 407.

[75] SHI F G. Size dependent thermal vibrations and melting in nanocrystals [J]. Journal of Materials Research, 1994, 9: 1307.

[76] DAVID T B, LEREAH Y, DEUTSCHER G, et al. Solid-liquid transition in ultra-fine lead particles [J]. Philosophical Magazine A, 1995, 71: 1135.

[77] JOHARI G P. Thermodynamic contributions from pre-melting or pre-transformation of finely dispersed crystals [J]. Philosophical Magazine A, 1998, 77: 1367.

[78] PETERS K F, COHEN J B, CHUNG Y W. Melting of Pb nanocrystals [J]. Physical Review B, 1998, 57: 13430.

[79] MORISHIGE K, KAWANO K. Freezing and melting of methyl chloride in a single cylindrical pore: Anomalous pore-size dependence of phase-transition temperature [J]. Journal of Physical Chemistry B, 1999, 103: 7906.

[80] JIANG Q, SHI H X, ZHAO M. Melting thermodynamics of organic nanocrystals [J]. Journal of Chemical Physics, 1999, 111: 2176.

[81] ZHANG Z, LI J C, JIANG Q. Modeling for size-dependent and dimension-dependent melting of nanocrystals [J]. Journal of Physics D: Applied Physics, 2000, 33: 2653.

[82] WEN Z, ZHAO M, JIANG Q. The melting temperature of molecular nanocrystals at the lower bound of mesoscopic size range [J]. Journal of Physics: Condensed Matter, 2000, 12: 8819.

[83] ZHANG Z, ZHAO M, JIANG Q. Melting temperature of semiconductor nanocrystals in the mesoscopic size range [J]. Semiconductor Science and Technology, 2001, 16: L33.

[84] PAWLOW P. Über die abnängikeit des schmelzpunktes von der oberflächenenergie eines festen kärpers [J]. Zeitschrift für Physikalische Chemie, 1909, 65: 545.

[85] HANSZEN K J. The melting points of small spherules [J]. Zeitschrift für Physik, 1960, 157: 523.

[86] JONES D R H. Review: The free energies of solid-liquid interfaces [J]. Journal of Materials Science, 1974, 9: 1.

[87] SAKA H, NISHIKAWA Y, IMURA T. Melting temperature of In particles embedded in an Al matrix [J]. Philosophical Magazine A, 1988, 57: 895.

[88] SHENG H W, REN G, PENG L M, et al. Melting process of nanometer-sized In particles embedded in an Al matrix synthesized by ball milling [J]. Journal of Materials Research, 1996, 11: 2841.

[89] CHATTOPADHYAY K, GOSWAMI R. Melting and superheating of metals and alloys [J]. Progress in Materials Science, 1997, 42: 287.

[90] ZHANG L, JIN Z H, ZHANG L H, et al. Superheating of confined Pb thin films [J]. Physical Review Letters, 2000, 85: 1484.

[91] JIANG Q, ZHANG Z, LI J C. Size-dependent superheating of nanocrystals embedded in matrix [J]. Chemical Physics Letters, 2000, 322: 549.

[92] JIANG Q, ZHANG Z, LI J C. Melting thermodynamics of nanocrystals embedded in matrix [J]. Acta Materialia, 2000, 48: 4791.

[93] JIANG Q, LIANG L H, LI J C. Thermodynamic superheating and relevant interface stability of low-dimensional metallic crystals [J]. Journal of Physics: Condensed Matter, 2001, 13: 565.

[94] REISS H, WILSON I B. The effects of size on melting point [J]. Journal of Colloid Science, 1948, 3: 551.

[95] ZHAO M, JIANG Q. Melting and surface melting of low-dimensional In crystals [J]. Solid State Communications, 2004, 130: 37 - 39.

[96] JIANG Q, SHI H X, ZHAO M. Free energy of crystal-liquid interface [J]. Acta Materialia, 1999, 47: 2109.

[97] WEN Z, ZHAO M, JIANG Q. Size range of solid-liquid interface free energy of organic crystals [J]. Journal of Physical Chemistry B, 2002, 106: 4266.

[98] JONES H. The solid-liquid interfacial energy of metals: Calculation versus measurements [J]. Materials Letters, 2002, 53: 364.

[99] MOTT N F. The resistance of liquid metals [J]. Proceedings of the Royal Society A, 1934, 146: 465.

[100] REGEL'A R, GLAZOV V M. Entropy of melting of semiconductors [J]. Semiconductors, 1995, 29: 405.

[101] FERMI E. Thermodynamics [M]. New York: Dover Publications, Inc. , 1963.

[102] PAPON P, LEBLOND J, MEIJER P H E. The physics of phase

transition: Concepts and applications [M]. Berlin: Springer, 2002.

[103] BEZRYADIN A, DEKKER C, SCHMID G. Electrostatic trapping of single conducting nanoparticles between nanoelectrodes [J]. Applied Physics Letters, 1997, 7: 1273 - 1275.

[104] BINDER K, FRATZL P. Phase transition in materials [M]. Weinheim: Wiley-VCH, 2001.

[105] KHACHATURYAN A G. Theory of structure phase transitions in solid [M]. New York: John Wiley & Sons, Inc., 1983.

[106] ARTEMEV A, WANG Y, KHACHATURYAN A G. Three-dimensional phase field model and simulation of martensitic transformation in multilayer systems under applied stresses [J]. Acta Materialia, 2000, 48: 2503 - 2518.

[107] SCHWARZ R B, KHACHATURYANM A G. Thermodynamics of open two-phase systems with coherent interfaces: Application to metal-hydrogen systems [J]. Acta Materialia, 2006, 54: 313 - 323.

[108] BRUCE A D, COWLEY R A. Structural Phase Transitions [M]. London: Taylor and Francis, 1981.

[109] LEVANYUK A P, SIGOV A S. Defects, structural phase transitions [M]. New York: Gordon and Breach, 1988.

[110] ABUDUKELIMU G, GUISBIERS G, WAUTELET M. Theoretical phase diagrams of nanowires [J]. Journal of Materials Research, 2006, 21: 2829 - 2834.

[111] CLARK S M, PRILLIMAN S G, ERDONMEZ C K, et al. Size dependence of the pressure-induced γ to α structural phase transition in iron oxide nanocrystals [J]. Nanotechnology, 2005, 16: 2813 - 2818.

[112] WANG C X, YANG G W. Thermodynamics of metastable phase nucleation at the nanoscale [J]. Materials Science and

Engineering, R: Reports, 2005, 49: 157 - 202.
[113] BARNARD A S. Using theory and modeling to investigate shape at the nanoscale [J]. Journal of Materials Chemistry, 2006, 16: 813 - 815.
[114] JIANG J Z. Phase transformations in nanocrystals [J]. Journal of Materials Science, 2004, 39: 5103 - 5110.
[115] RONG Y H. Phase transformations and phase stability in nanocrystalline materials [J]. Current Opinion in Solid State & Materials Science, 2005, 9: 287 - 295.
[116] NAVROTSKY A. Energetics of nanoparticle oxide: Interplay between surface energy and polymorphism [J]. Geochemical Transactions, 2003, 4: 34 - 37.
[117] BARNARD A S, YEREDLA R R, XU H F. Modeling the effect of particle shape on the phase stability of ZrO_2 nanoparticles [J]. Nanotechnology, 2006, 17: 3039 - 3047.
[118] WANG C X, HIRANO M, HOSONO H. Origin of Diameter-dependent growth direction of silicon nanowires [J]. Nano Letters, 2006, 6: 1552 - 1555.
[119] HILL T L. Thermodynamics of small systems (I) [M]. New York: Dover Publications, Znc., 1963.
[120] HILL T L. Thermodynamics of small systems (II) [M]. New York: Dover Publications, Znc., 1964.
[121] TSALLIS C. Possible generalization of Boltzmann-Gibbs statistics [J]. Journal of Statistical Physics, 1988, 52: 479 - 487.
[122] RAJAGOPAL A K. Von neumann and Tsallis entropies associated with the Gentile interpolative quantum statistics [J]. Physics Letters A, 1996, 214: 127 - 130.
[123] YANG C C, LI J C, JIANG Q. Effects of pressure on melting temperature of silicon determined by Clapeyron equation [J]. Chemical Physics Letters, 2003, 372: 156 - 159.

[124] GARVIE R C. The occurrence of metastable tetragonal zirconia as crystallite size effect [J]. The Journal of Physical Chemistry, 1965, 69: 1238 – 1243.

[125] GARVIE R C, HANNINK R H, PASCOE R T. Ceramic steel [J]. Nature, 1975, 258: 703 – 704.

[126] GARVIE R C. Stabilization of the tetragonal structure in Zirconia microcrystals [J]. The Journal of Physical Chemistry, 1978, 82: 218 – 223.

[127] ZHANG H Z, BANFIELD J F. Understanding polymorphic phase transformation behavior during growth of nanocrystalline aggregates: Insights from TiO_2 [J]. Journal of Physical Chemistry B, 2000, 104: 3481 – 3487.

[128] ZHANG H Z, BANFIELD J F. Size dependent of the kinetic rate constant for phase transformation in TiO_2 nanoparticles [J]. Chemistry of Materials, 2005, 17: 3421 – 3425.

[129] HUANG F, BANFIELD J F. Size-dependent phase transformation kinetics in nanocrystalline ZnS [J]. Journal of the American Chemical Society, 2005, 127: 4523 – 4529.

[130] ZHANG H Z, HUANG F, GILBERT B, et al. Molecular dynamics simulations, thermodynamics analysis, and experimental study of phase stability of zinc sulfide nanoparticles [J]. Journal of Physical Chemistry B, 2003, 107: 13051 – 13060.

[131] BARNARD A S, ZAPOL P. A model for the phase stability of arbitrary nanoparticles as a function of size and shape [J]. Journal of Chemical Physics, 2004, 121: 4276 – 4283.

[132] BARNARD A S, ZAPOL P. Predicting the energetics, phase stability, and morphology evolution of faceted and spherical anatase nanocrystals [J]. Journal of Physical Chemistry B, 2004, 108: 18435 – 18440.

[133] BARNARD A S, CURTISS L A. Prediction of TiO_2 nanoparticles

phase and shape transitions controlled by surface chemistry[J]. Nano Letters, 2005, 5: 1261 -1266.

[134] BARNARD A S, CURTISS L A. Computational nano-morphology: Modeling shape as well as size [J]. Reviews on Advanced Materials Science, 2005, 10: 105 -109.

[135] BARNARD A S, XIAO Y, CAI Z. Modeling the shape and orientation of ZnO nanobelts [J]. Chemical Physics Letters, 2006, 419: 313 -316.

[136] FAN H J, BARNARD A S, ZACHARIAS M. ZnO nanowires and nanobelts: Shape section and thermodynamic modeling[J]. Applied Physics Letters, 2007, 90: 143116 (1 -3).

[137] YANG C X, YANG Y H, XU N S, et al. Thermodynamics of diamond nucleation on the nanoscale [J]. Journal of the American Chemical Society, 2004, 126: 11303 -11306.

[138] LIU Q X, WANG C X, LI S W, et al. Homogeneous nucleation of diamond in the gas phase: A nano-scale thermodynamic approach [J]. Carbon, 2004, 42: 585 -590.

[139] ZHANG C Y, WANG C X, YANG Y H, et al. Nucleation thermodynamics of cubic boron nitride upon high-pressure and high-temperature supercritical fluid system in nanoscale [J]. Journal of Physical Chemistry B, 2004, 108: 2589 -2593.

[140] WANG C X, YANG Y H, YANG G W. Nanothermodynamic analysis of the low-threshold-pressure-synthesized cubic boron nitride in supercritical-fluid systems [J]. Applied Physics Letters, 2004, 84: 3034 -3036.

[141] WANG B, YANG Y H, XU N S, et al. Mechanisms of size-dependent shape evolution of one-dimensional nanostructure growth [J]. Physical Review B, 2006, 74: 235305 (1 -6).

[142] WANG C X, WANG B, YANG Y H, et al. Thermodynamic and kinetic size limit of nanowire growth[J]. Journal of Physical

Chemistry B, 2005, 109: 9966 – 9969.

[143] ZHANG H Z, BANFIELD J F. Thermodynamic analysis of phase stability of nanocrystalline titania [J]. Journal of Materials Chemistry, 1998, 8: 2073 – 2076.

[144] BLAKELY J M. Introduction to the properties of crystal surfaces [M]. 1st ed. Oxford: Pergamon Press, 1973: 7 – 10.

[145] GIBBS J W. The scientific papers of J. Willard Gibbs VI. Thermodynamics [M]. London: Longmans-Green, 1906: 55.

[146] CAHN J W. Surface stress and the chemical equilibrium of small crystals—I. The case of the isotropic surface [J]. Acta Metallurgica, 1980, 28: 1333 – 1338.

[147] YANG B, ASTA M, MRYASOV O N, et al. Equilibrium Monte Carlo simulations of A1-L1$_0$ ordering in FePt nanoparticles [J]. Scripta Materialia, 2005, 53: 417 – 422.

[148] DELOGU F. The mechanism of chemical disordering in Cu$_3$Au nanometer-sized systems [J]. Nanotechnology, 2007, 18: 235706 (8pp).

[149] SATO K, HIROTSU Y. Structure and magnetic property changes of epitaxially grown L1$_0$-FePd isolated nanoparticles on annealing [J]. Journal of Applied Physics, 2003, 93: 6291 – 6298.

[150] KANG S S, JIA Z Y, NIKLES D E, et al. Synthesis and phase transition of self-assembled FePd and FePdPt nanoparticles[J]. Journal of Applied Physics, 2004, 95: 6744 – 6746.

[151] SATO K, KONNO T J, HIROTSU Y. L1$_0$-type ordered structure of FePd nanoparticles studied by high-resolution transmission electron microscopy [J]. Journal of Materials Science, 2008, 2: 619 – 620.

[152] KEITA W, KURA H, SATO T. Transformation to L1$_0$ structure in FePd nanoparticles synthesized by modified polyol process[J].

Science and Technology of Advanced Materials, 2006, 7: 145 -149.

[153] ROSSI G, FERRANDO R, MOTTET C. Structure and chemical ordering in CoPt nanoalloys [J]. Faraday Discussions, 2008, 138: 193 -210.

[154] GAMBARDELLA P, RUSPONI S, VERONESE M. Giant magnetic anisotropy of single Cobalt atoms and nanoparticles [J]. Science, 2003, 300: 1130 -1133.

[155] KIM J S, KOO Y M, LEE B J. Modified embedded-atom method interatomic potential for the FePt alloy system [J]. Journal of Materials Research, 2006, 21: 199 -208.

[156] LU H M, CAO Z H, ZHAO C L, et al. Size-dependent ordering and Curie temperatures of FePt nanoparticles [J]. Journal of Applied Physics, 2008, 103: 123526.

[157] YANG C C, XIAO M X, LI W, et al. Size effects on Debye temperature, Einstein temperature, and volume thermal expansion coefficient of nanocrystals [J]. Solid State Communications, 2006, 139: 148 -152.

[158] YANG C C, JIANG Q. Size effect on the phase stability of nanostructures [J]. Current Nanoscience, 2008, 4: 179 -200.

[159] LU H M, MENG X K. Size dependence of Magnetostructural transition in MnBi nanorods [J]. Journal of Physical Chemistry C, 2010, 114: 2932 -2935.

[160] DEVOLDER T, PIZZINI S, VOGEL J, et al. X-ray absorption analysis of sputter-grown Co/Pt stackings before and after helium irradiation[J]. European Physical Journal B,2001,22:193 -201.

[161] http://www.webelements.com（元素周期表）.

[162] CHEPULSKII R V, BUTLER W H. Temperature and particle-size dependence of the equilibrium order parameter of FePt al-

loys [J]. Physical Review B, 2005, 72: 134205.

[163] CHINNASAMY C N, JEYADEVAN B, SHINODA K, et al. Polyol-process-derived CoPt nanoparticles: Structural and magnetic properties [J]. Journal of Applied Physics, 2003, 93: 7583.

[164] ALLOYEAU D, RICOLLEAU C, OILAWA T, et al. STEM nanodiffraction technique for structural analysis of CoPt nanoparticles [J]. Ultramicroscopy, 2008, 108: 656 – 662.

[165] LIU D, ZHU Y F, JIANG Q. Site-and Structure-dependent cohesive energy in several Ag clusters [J]. Journal of Physical Chemistry C, 2009, 113: 10907 – 10912.

[166] GRAF C, VAN BLAADEREN A. Metallodielectric Colloidal Core-Shell Particles for Photonic Applications [J]. Langmuir, 2002, 18: 524.

[167] FERNANDEZ J L, WALSH D A, BARD A J. Thermodynamic guidelines for the design of bimetallic catalysts for oxygen electroreduction and rapid screening by scanning electrochemical microscopy. M – Co (M: Pd, Ag, Au) [J]. Journal of the American Chemical Society, 2005, 127: 357.

[168] BRONSTEIN L M, CHERNYSHOV D M, VOLKOV I O. Structure and properties of bimetallic colloids formed in Polystyrene-block-Poly-4-vinylpyridine micelles: Catalytic behavior in selective Hydrogenation of dehydrolinalool [J]. Journal of Catalysis, 2000, 196: 302.

[169] THOMAS J M, RAJA R, JOHNSON B F G, et al. Bimetallic catalysts and their relevance to the Hydrogen economy [J]. Industrial & Engineering Chemistry Research, 2003, 42: 1563.

[170] CHUSHAK Y G, BARTELL L S. Freezing of Ni-Al Bimetallic Nanoclusters in Computer Simulations [J]. Journal of Physical Chemistry B, 2003, 107: 3747.

[171] JEONG U, KIM J U, XIA Y N. Monodispersed spherical colloids of Se@CdSe: Synthesis and use as building blocks in fabricating photonic crystals [J]. Nano Letters, 2005, 5: 937.

[172] MAYE M M, KARIUKI N N, LUO J, et al. Electrocatalytic reduction of Oxygen: Gold and gold-platinum nanoparticle catalysts prepared by two-phase protocol [J]. Gold Bulletin, 2004, 37: 217.

[173] LUO Y, MAYE M M, HAN L, et al. Gold-platinum alloy nanoparticle assembly as catalyst for methanol electrooxidation [J]. Chemical Communications, 2001 (5): 473.

[174] ZHONG C J, LUO J, NJOKI P N, et al. Fuel cell technology: Nano-engineered multimetallic catalysts [J]. Energy & Environmental Science, 2008, 1: 454.

[175] ZHONG C J, LUO J, FANG B, et al. Nanostructured catalysts in fuel cells [J]. Nanotechnology, 2010, 21: 062001.

[176] CAO L Y, TONG L M, DIAO P, et al. Kinetically controlled Pt deposition onto self-assembled Au colloids: Preparation of Au (Core) -Pt (Shell) nanoparticle assemblies [J]. Chemistry of Materials, 2004, 16: 3239.

[177] TADA H, SUZUKI F, ITO S, et al. Adsorption of 2, 2′-dipyridyl disulfide on Au/Pt core/shell bimetallic clusters loaded on TiO_2: Fine control of adsorptivity for organosulfur compounds[J]. ChemPhysChem, 2002, 3: 617.

[178] NAKANISHI M, TAKATANI H, KOBAYASHI Y, et al. Characterization of binary gold/platinum nanoparticles prepared by sonochemistry technique [J]. Applied Surface Science, 2005, 241: 209.

[179] YANG Z, YANG X N, XU Z J. Molecular dynamics simulation of the melting behavior of Pt-Au nanoparticles with Core-Shell structure [J]. Journal of Physical Chemistry C, 2008,

112: 4937.
[180] WANG F, LIU P, ZHANG D J. Structures of Au/Pt bimetallic clusters: Homogeneous or segregated? [J]. Journal of Molecular Modeling, 2011, 17: 1069.
[181] LIU H B, PAL U, ASCENCIO J A. Thermodynamic stability and melting mechanism of bimetallic Au-Pt nanoparticles [J]. Journal of Physical Chemistry C, 2008, 112: 19173.
[182] SHI R W, SHAO J L, ZHU X L, et al. On the melting and freezing of Au-Pt nanoparticles confined in single-walled carbon nanotubes [J]. Journal of Physical Chemistry C, 2011, 115: 2961.
[183] QI W H, LEE S T. Phase stability, melting, and alloy formation of Au-Ag bimetallic nanoparticles [J]. Journal of Physical Chemistry C, 2010, 114: 9580.
[184] CHEN F Y, CURLEY B C, ROSSI G, et al. Structure, melting, and thermal stability of 55 atom Ag-Au nanoalloys [J]. Journal of Physical Chemistry C, 2007, 111: 9157.
[185] KIM D H, KIM H Y, RYU J H, et al. Phase diagram of Ag-Pd bimetallic nanoclusters by molecular dynamics simulations: Solid-to-liquid transition and size-dependent behavior [J]. Physical, Chemistry Chemical Physics, 2009, 11: 5079.
[186] SANKARANARAYANAN S K R S, BHETHANABOTLA V R, JOSEPH B. Molecular dynamics simulation study of the melting of Pd-Pt nanoclusters [J]. Physical Review B, 2005, 71: 195415.
[187] LI G J, WANG Q, LI D G, et al. Size and composition effects on the melting of bimetallic Cu-Ni clusters studied via molecular dynamics simulation [J]. Materials Chemistry and Physics, 2009, 114: 746.
[188] SHANDIZ M A, SAFAEI A, SANJABI S, et al. Modeling size de-

pendence of melting temperature of metallic nanoparticles [J]. Journal of Physics and Chemistry Solids, 2007, 68: 1396.

[189] NANDA K K, SAHU S N, BEHERA S N. Liquid-drop model for the size-dependent melting of low-dimensional systems [J]. Physical Review A, 2002, 66: 013208.

[190] QI W H, WANG M P. Size and shape dependent melting temperature of metallic nanoparticles [J]. Materials Chemistry and Physics, 2004, 88: 280.

[191] LEWIS L J, JENSEN P, BARRAT J L. Melting, freezing, and coalescence of gold nanoclusters [J]. Physical Review B, 1997, 56: 2248.

[192] SUN C Q, WANG Y, TAY B K, et al. Correlation between the melting point of a nanosolid and the cohesive energy of a surface atom [J]. Journal of Physical Chemistry B, 2002, 106: 10701.

[193] LU H M, LI P Y, CAO Z H, et al. Size-, shape-, and dimensionality-dependent melting temperatures of nanocrystals [J]. Journal of Physical Chemistry C, 2009, 113: 7598.

[194] YANG C C, LI S. Size, dimensionality, and constituent stoichiometry dependence of bandgap energies in semiconductor quantum dots and wires [J]. Journal of Physical Chemistry C, 2008, 112: 2851.

[195] YANG C C, LI S. Size-, dimensionality-, and composition-dependent Debye temperature of monometallic and bimetallic nanocrystals in the deep nanometer scale [J]. Physica Status Solidi B, 2011, 248: 1375.

[196] FOX T G. Influence of diluent and of copolymer composition on the glass temperature of a polymer system [J]. Bulletin of the American Physical Society, 1956, 1: 123.

[197] WANJALA B N, LUO J, FANG B, et al. Gold-platinum nano-

particles: Alloying and phase segregation [J]. Journal of Materials Chemistry, 2011, 21: 4012.

[198] JACKSON C L, MCKENNA G B. The melting behavior of organic materials confined in porous solids [J]. Journal of Chemical Physics, 1990, 93: 9002.

[199] HANSEN E W, STÖCKER M, SCHMIDT R. Low-temperature phase transition of water confined in mesopores probed by NMR. Influence on pore size distribution [J]. The Journal of Physical Chemistry, 1996, 100: 2195.

[200] JIANG Q, ZHANG S, ZHAO M. Size-dependent melting point of noble metals [J]. Materials Chemistry and Physics, 2003, 82: 225 - 227.

[201] ALKESKJOLD T T, SCOLARI L, NOORDEGRAAF D, et al. Integrating liquid crystal based optical devices in photonic crystal fibers [J]. Optical and Quantum Electronics, 2007, 39: 1009 - 1019.

[202] SHAH R R, ABBOTT N L. Using liquid crystals to image reactants and products of acid-base reactions on surface with micrometer resolution [J]. Journal of the American Chemical Society, 1999, 121: 11300 - 11310.

[203] BRAKE J M, DASCHNER M K, LUK Y Y et al. Biomolecular interactions at phospholipid-decorated surfaces of liquid crystals[J]. Science, 1997, 276: 1533 - 1536.

[204] SHAO Y, ZERDA T W. Phase transitions of liquid crystal PAA in confined geometries [J]. Journal of Physical Chemistry B, 1998, 102: 3387 - 3394.

[205] STANNARIUS R, KREMER F. Liquid crystals in confining geo-metries [J]. Lecture Notes in Physics, 2004, 634: 301 - 336.

[206] GRIGORIADIS C, DURAN H, STEINHART M, et al. Sup-

pression of phase transitions in a confined rodlike liquid crystal [J]. ACS Nano, 2011, 5: 9208 - 9215.

[207] CRAMER C H, CRAMER T H, KREMER F, et al. Measurement of orientational order and mobility of a nematic liquid crystal in random nanometer confinement [J]. Journal of Chemical Physics, 1997, 106: 3730.

[208] IANNACCHIONE G S, FINOTELLO D. Calorimetric study of phase transitions in confined liquid crystals [J]. Physical Review Letters, 1992, 69: 2094.

[209] KRALJ S, ZIDANŠEK A, LAHAJNAR G, et al. Phase behaviour of liquid crystals confined to controlled porous glass studied by deuteron NMR [J]. Physical Review E, 1998, 57: 3021.

[210] DADMUN M D, MUTHUKUMAR M. The nematic to isotropic transition of a liquid crystal in porous media [J]. Journal of Chemical Physics, 1993, 98: 4850.

[211] VILFAN M, APIH T, GREGOROVIČ A, et al. Surface-induced order and diffusion in 5CB liquid crystal confined to porous glass [J]. Magnetic Resonance Imaging, 2001, 19: 433 - 438.

[212] GRINBERG F, AND KIMMICH R. Pore size dependence of the dipolar-correlation effect on the stimulated echo in liquid crystals confined in porous glass [J]. Journal of Chemical Physics, 1996, 105: 3301.

[213] LANG X Y, JIANG Q. Size effect on glass transition temperature of nanopolymers [J]. Solid State Phenomena, 2007, 121 - 123: 1317.

[214] THOTE A J, GUPTA R B. Hydrogen-bonding effects in liquid crystals for application to LCDs [J]. Industrial & Engineering Chemistry Research, 2003, 42: 1129 - 1136.

[215] TEWARI R P, KHAN G, SHUKLA A, et al. Comparative study of Azoxy-based liquid crystal (p-azoxyanisole, p-azoxyphenetole, ethyl-p-azoxybenzoate, ethyl-p-azoxycinnamate and n-octyl-p-azoxycinnamate) based on partial atomic charges [J]. Archive of Physics Research, 2012, 3 (3): 175 – 191.

[216] JIANG Q, LIANG L H, ZHAO M. Modeling of the melting temperature of nano-ice in MCM-41 pores [J]. Journal of Physics: Condensed Matter, 2001, 13: L397 – L401.

[217] KUZE N, EBIZUKA M, FUJIWARA H, et al. Molecular structure of p-azoxyanisole, a mesogen, determined by gas-phase electron diffraction augmented by ab initio calculations [J]. The Journal of Physical Chemistry A, 1998, 102: 2080.

[218] CLARK S J, ADAM C J, ACKLAND G J, et al. Properties of liquid crystal molecules from first principles computer simulation [J]. Liquid Crystals, 1997, 22: 469 – 475.

[219] JAISWAL A K. Solid polymorphs of p-azoxyanisole (PAA) [J]. National Academy Science Letters (India), 1982, 5: 23.

[220] HORIUCHI K, YAMAMURA Y, PELKA R, et al. Entropic contribution of flexible terminals to mesophase formation revealed by thermodynamic analysis of 4-alkyl-4′-isothiocyanato-biphenyl (nTCB) [J]. Journal of Physical Chemistry B, 2010, 114: 4870 – 4875.

[221] KING P D C, VEAL T D, FUCHS F, et al. Band gap, electronic structure, and surface electron accumulation of cubic and rhombohedral In_2O_3 [J]. Physical Review B, 2009, 79: 205211.

[222] LI X, WANLASS M W, GESSERT T A, et al. High-efficiency indium tin oxide/indium phosphide solar cells [J]. Applied Physics Letters, 1989, 54: 2674.

[223] WANG D, ZOU Z, YE J. Photocatalytic water splitting with the

Cr-doped $Ba_2In_2O_5/In_2O_3$ composite oxide semiconductors [J]. Chemistry of Materials, 2005, 17: 3255.

[224] NGUYEN P, NG H T, YAMADA T, et al. Direct integration of metal oxide nanowire in vertical field-effect transistor [J]. Nano Letters, 2004, 4: 651.

[225] SHANNON R D. New high pressure phases having the corundum structure [J]. Solid State Communications, 1966, 4: 629.

[226] REID A F, RINGWOOD A E. High-pressure scandium oxide and its place in the molar volume relationships of dense structures of M_2X_3 and ABX_3 type[J]. Journal of Geophysical Research, 1969, 74: 3238.

[227] ATOU T, KUSABA K, FUKUOKA K, et al. Shock-induced phase transition of M_2O_3 (M = Sc, Y, Sm, Gd, and In) -type compounds[J]. Journal of Solid State Chemistry, 1990, 89: 378.

[228] XU J Q, CHEN Y P, PAN Q Y, et al. A new route for preparing corundum-type In_2O_3 nanorods used as gas-sensing materials[J]. Nanotechnology, 2007, 18: 115615.

[229] GURLO A. Nanosensors: Towards morphological control of gas sensing activity, SnO_2, In_2O_3, ZnO, and WO, case studies[J]. Nanoscale, 2011, 3: 154.

[230] EPIFANI M, SICILIANO P, GURLO A, et al. Ambient pressure synthesis of corundum-type In_2O_3 [J]. Journal of the American Chemical Society, 2004, 126: 4078.

[231] GURLO A, LAUTERBACH S, MIEHE G, et al. Nanocubes or nanorhombohedra? Unusual crystal shapes of corundum-type indium oxide [J]. Journal of Physical Chemistry C, 2008, 112: 9209.

[232] FARVID S S, DAVE N, RADOVANOVIC P V. Phase-Con-

trolled synthesis of colloidal In_2O_3 nanocrystals via size-structure correlation[J]. Chemistry of Materials, 2010, 22: 9.

[233] GURLO A. Structural stability of high-pressure polymorphs in In_2O_3 nanocrystals: Evidence of stress-induced transition? [J]. Angewandte Chemie International Edition, 2010, 49: 5610.

[234] QI J, LIU J F, HE Y, et al. Compression behavior and phase transition of cubic In_2O_3 nanocrystals[J]. Journal of Applied Physics, 2011, 109: 063520.

[235] GARCIA-DOMENE B, ORTIZ H M, GOMIS O. High-pressure lattice dynamical study of bulk and nanocrystalline In_2O_3[J]. Journal of Applied Physics, 2012, 112: 123511.

[236] BARNARD A S, XU H F. An environmentally sensitive phase map of titania nanocrystals[J]. ACS Nano, 2008, 2: 2237.

[237] CHEN Z W, SUN C Q, ZHOU Y C, et al. Size dependence of the pressure-induced phase transition in nanocrystals[J]. Journal of Physical Chemistry C, 2008, 112: 2423.

[238] SUN C Q. Dominance of broken bonds and nonbonding electrons at the nanoscale[J]. Nanoscale, 2010, 2: 1930.

[239] LI S, ZHENG W T, JIANG Q. Size and pressure effects on solid transition temperatures of ZrO_2[J]. Scripta Materialia, 2006, 54: 2091.

[240] LI S, WEN Z, JIANG Q. Pressure-induced phase transition of CdSe and ZnO nanocrystals[J]. Scripta Materialia, 2008, 59: 526.

[241] YANG C C, LI S. Size-dependent phase stability of silver nanocrystals[J]. The Journal of Physical Chemistry C, 2008, 112: 16400.

[242] DELLEY B. An all-electron numerical method for solving the local density functional for polyatomic molecules[J]. Journal of

Chemical Physics, 1990, 92: 508.

[243] DELLEY B. From molecules to solids with the DMol^3 approach[J]. Journal of Chemical Physics, 2000, 113: 7756.

[244] HAMMER B, HANSEN L B, NORSKOV J K. Improved adsorption energetics within density-functional theory using revised Perdew-Burke-Ernzerhof functionals[J]. Physical Review B, 1999, 59: 7413.

[245] LIU W, ZHENG W T, JIANG Q. First-principles study of the surface energy and work function of Ⅲ-Ⅴ semiconductor compounds[J]. Physical Review B, 2007, 75: 235322.

[246] BOETTGER J C. Nonconvergence of surface energies obtained from thin film calculations[J]. Physical Review B, 1994, 49: 16798.

[247] WALSH A, RICHARD C, CATLOW A. Structure, stability and work functions of the low index surfaces of pure indium oxide and Sn-doped indium oxide (ITO) from density functional theory[J]. Journal of Materials Chemistry, 2010, 20: 10438.

[248] KELVIN H, ZHANG L, WALSH A, et al. Surface energies control the self-organization of oriented In_2O_3 nanostructures on cubic Zirconia[J]. Nano Letters, 2010, 10: 3740.

[249] LIDE D R. Handbook of Chemistry and Physics [M]. 87th ed. UK: Taylor and Francis Group, Boca Raton, Fl, 2007.

[250] PREWITT C T, SHANNON R D, ROGERS D B, et al. C rare earth oxide-corundum transition and crystal chemistry of oxides having the corundum structure [J]. Inorganic Chemistry, 1969, 8: 1985 – 1993.

[251] GURLO A, KROLL P, RIEDEL R. Metastability of Corundum-type In_2O_3 [J]. Chemistry-A European Journal, 2008, 14: 3306.

[252] ZACHARIASEN W. The crystal structure of the modification C

of the sesquioxides of the rare earth metals, and of indium and thallium [J]. Norsk Geologisk Tidsskrift, 1927, 3: 310.

[253] LIU D, LEI W W, ZOU B, et al. High-pressure X-ray diffraction and Raman spectra study of indium oxide[J]. Journal of Applied Physics, 2008, 104: 083506.

[254] SHU S W, YU D B, WANG Y, et al. Thermal-induced phase transition and assembly of hexagonal metastable In_2O_3 nanocrystals: A new approach to In_2O_3 functional materials[J]. Journal of Crystal Growth, 2010, 312: 3111.

[255] ZHU Y F, LIAN J S, JIANG Q. Modeling of the melting point, debye temperature, thermal expansion coefficient, and the specific heat of nanostructured materials[J]. Journal of Physical Chemistry C, 2009, 113: 16896-16900.

[256] CHU D W, ZENG Y P, JIANG D L, et al. Tuning the phase and morphology of In_2O_3 nanocrystals via simple solution routes[J]. Nanotechnology, 2007, 18: 435605.

[257] ZAPIEN J A, JIANG Y, MENG X M, et al. Room-temperature single nanoribbon lasers[J]. Applied Physics Letters, 2004, 84: 1189.

[258] PAN Y W S C, Dong S H, et al. An investigation on the pressure-induced phase transition of nanocrystalline ZnS[J]. Journal of Physics: Codensed Matter, 2002, 14: 10487.

[259] TOLBERT S H, ALIVISATOS A P. The wurtzite to rock salt structure transformation in CdSe nanocrystals under high pressure[J]. Journal of Chemical Physics, 1995, 102: 4642.

[260] KUMAR R S, CORMELIUS A L, NICOL M F. Structure of nanocrystalline ZnO up to 85 GPa[J]. Current Applied Physics, 2007, 7: 135.

[261] JIANG J Z, OLSEN J S, GERWARD L, et al. Structural stability in nanocrystalline ZnO[J]. Europhysics Letters, 2000,

50: 48.
[262] GRZANKA E, GIERLOTKA S, STELMAKH S, et al. Phase transition in nanocrystalline ZnO[J]. Zeitschrift für Kristallographie Supplement, 2006, 23: 337.
[263] WANG Z W, SAXENA S K, PISCHEDDA V, et al. In situ X-ray diffraction study of the pressure-induced phase transformation in nanocrystalline CeO_2[J]. Physical Review B, 2001, 64: 012102.
[264] WANG Z W, TAIT K, ZHAO Y S, et al. Size-induced reduction of transition pressure and enhancement of bulk modulus of AlN nanocrystals[J]. The Journal of Physical Chemistry B, 2004, 108: 11506 – 11508.
[265] SHUKLA S, SEAL S. Mechanism of room temperature metastable tetragonal phase stabilization in zirconia[J]. International Materials Reviews, 2005, 50: 45 – 64.
[266] JIANG J Z, OLSEN S S, GERWARD L. Grain-size and alloying effects on the pressure-induced bcc-to-hcp transition in nanocrystalline iron[J]. Materials Transactions Online — The Japan Institute of Metals, 2001, 42: 1571.
[267] JIANG J Z, GERWARD L, OLSEN J S. Pressure induced phase transformation in nanocrystal[J]. SnO_2, Scripta Materialia, 2001, 44: 1983.
[268] JIANG X B, JIANG M, ZHAO M. Thermodynamic Understanding of Phase Transitions of In_2O_3 Nanocrystals[J]. Chemical Physical Letters, 2013, 563: 76 – 79.
[269] HE Y, LIU J F, CHEN W, et al. High-pressure behavior of SnO_2 nanocrystals[J]. Physical Review B, 2005, 72: 212102 – 4.
[270] LI S, LIAN J S, JIANG Q. Modeling size and surface effects on ZnS phase selection[J]. Chemical Physical Letters, 2008, 455: 202 – 206.

[271] ZHAO D S, ZHAO M, JIANG Q. Size and temperature dependence of nanodiamond-nanographite transition related with surface stress [J]. Diamond and Related Materials, 2002, 11: 234.

[272] LI S, JIANG Z H, JIANG Q. Thermodynamic phase stability of three nano-oxide[J]. Materials Research Bulletin, 2008, 43: 3149-3154.

[273] WINTERER M, NITSCHE R, REDFERN S A T, et al. Phase stability in nanostructured and coarse grained zirconia at high pressure[J]. Nanostructured Materials, 1995, 6: 679-688.

[274] BLOCK S, DA JORNADA J A H, PIERMARINI G J. Pressure-temperature phase diagram of zirconia[J]. Journal of the American Ceramic Society, 1985, 68: 497-499.

[275] DA JORNADA J A H, PIERMARINI G J, BLOCK S. Metastable retention of a high-pressure phase of zirconia[J]. Journal of the American Ceramic Society, 1987, 70: 628-630.

[276] JIANG Q, LU H M. Size dependent interface energy and its applications[J]. Surface Science Reports, 2008, 53: 427-464.

[277] TOLMAN R C. The effect of droplet size on surface tension[J]. Journal of Chemical Physics, 1949, 17: 333-337.

[278] JAMSHIDIAN M, THAMBURAJA P, RABCZUK T. A continuum state variable theory to model the size-dependent surface energy of nanostructures [J]. Physical Chemistry Chemical Physics, 2015, 17: 25494-25498.

[279] AQRA F, AYYAD A. Surface free energy of alkali and transition metal nanoparticles [J]. Applied Surface Science, 2014, 314: 308-313.

[280] ABDUL-HAFIDH E H, Aïssa B. A theoretical prediction of the paradoxical surface free energy for FCC metallic nanosolids[J]. Applied Surface Science, 2016, 379: 411-414.

[281] ZHANG S. The relationship between the size-dependent melting point of nanoparticles and the bond number [J]. Nanomaterials and Energy, 2016, 5: 125 –131.

[282] NANDA K K, KRUIS F E, Fissan H, et al. Higher surface energy of free nanoparticles [J]. Plysical Review Letters, 2003, 91: 106102.

[283] Medasani B, Park Y H, Vasiliev I. Theoretical study of the surface energy, stress, and lattice contraction of silver nanoparticles [J]. Physical Review B, 2007, 75: 235436.

[284] ZHAO M, ZHENG W T, Li J C, et al. Atomistic origin, temperature dependence, and responsibilities of surface energetics: An extended broken-bond rule [J]. Physical Review B, 2007, 75: 085427.

[285] TAKRORI F M , AYYAD A. Surface energy of metal alloy nanoparticles [J]. Applied Surface Science, 2017, 401: 65 –68.

[286] Liang L H, Zhao M, Jiang Q. Melting enthalpy depression of nanocrystals based on surface effect [J]. Journal of Materials Science letters, 2002, 21: 1843 –1845.

[287] Bartell L S. Tolman's δ, surface curvature, compressibility effects, and the free energy of drops [J]. Journal of Physical Chemistry B, 2001, 105: 11615 –11618.

[288] Barnard A S, Curtiss L A . Predicting the shape and structure of face-centered cubic gold nanocrystals smaller than 3 nm [J]. Chemistry Physical Chemical Physics, 2006, 7: 1544 – 1553.

[289] Samsonov V M, Shcherbakov L M , Novoselov A R, et al. Investigation of the microdrop surface tension and the linear tension of the wetting perimeter on the basis of similarity concepts and the thermodynamic perturbation theory [J]. Colloids and Surfaces A – Physicchemical and Engineering Aspects, 1999,

160: 117-121.

[290] Shandiz M A. Effective coordination number model for the size dependency of physical properties of nanocrystals [J]. Journal of Physics - Condensed Matter, 2008, 20: 325237.

[291] JIA M, LAI Y Q, Tian Z L, et al. Calculation of the surface free energy of fcc copper nanoparticles [J]. Modelling and Simulation in Materials Science and Engineering, 2009, 17: 015006.

[292] OUYANG G, TAN X, Cai M Q, et al. Surface energy and shrinkage of a nanocavity [J]. Applied Plysics Letters, 2006, 89: 183104.

[293] WACHOWICZ E, KIEJNA A. Bulk and surface properties of hexagonal-close-packed Be and Mg [J]. Journal of Physics - Condensed Matter, 2001, 13: 10767.

[294] KIEJNA A, PEISERT J, Scharoch P. Quantum-size effect in thin Al (110) slabs [J]. Surface Science, 1999, 432: 54-60.

[295] LUO W H, HU W Y, Su K L, et al. The calculation of surface free energy based on embedded atom method for solid nickel[J]. Applied Surface Science 2013, 265: 375-378.

[296] PLIETH W J. The work function of small metal particles and its relation to electrochemical properties [J]. Surface Science, 1985, 156: 530-535.

[297] SHIM J H, LEE B J, CHO Y W. Thermal stability of unsupported gold nanoparticle: A molecular dynamics study [J]. Surface Science, 2002, 512: 262-268.

[298] OUYANG G, WANG C X, YANG G W. Surface energy of nanostructural materials with negative curvature and related size effects [J]. Chemical Reviews, 2009, 109: 4221-4247.

[299] VERWEY E W J. Electronic conduction of magnetite (Fe_3O_4) and its transition point at low temperature[J]. Nature, 1939,

144: 327 – 328.
[300] SENN M S, WRIGHT J P, ATTFIELD J P. Charge order and three-site distortions in the Verwey structure of magnetite[J]. Nature, 2012, 481: 173 – 176.
[301] GOYA G F, BERQUO T S, FONSECA F C, et al. Static and dynamic magnetic properties of spherical magnetite nanopartilces[J]. Journal of Applied Physics, 2003, 94: 3520 – 3528.
[302] SANTOYO SALAZAR J, PEREZ L, DE ABRIL O, et al. Magnetic iron oxide nanoparticles in 10 – 40 nm range: Composition in terms of magnetite/maghemite ratio and effect on the magnetic properties[J]. Chemistry of Materials, 2001, 23: 1379 – 1386.
[303] PODDAR P, FRIED T, MARKOVICH G. First-order metal-insulator transition and spin-polarized tunneling in Fe_3O_4 nanocrystals[J]. Physical Review B, 2002, 65: 172405.
[304] SNOW C L, SHI Q, BOERIO-GOATES J, et al. Heat capacity studies of nanocrystalline magnetite (Fe_3O_4) [J]. Journal of Physical Chemistry C, 2010, 114: 21100 – 21108.
[305] LEE J, KWON S G, PARK J G, et al. Size dependent of metal-insulator transition in stoichiometric Fe_3O_4 nanocrystals[J]. Nano letters, 2015, 15 (7): 4337 – 4342.
[306] HEVRONI A, BAPNA M, PIOTROWSKI S, et al. Tracking the Verwey transition in single magnetite nanocrystals by variable-temperature scanning tunneling microscopy[J], The Journal of Physical Chemistry Letters, 2016, 7: 1661 – 1666.
[307] MOHAPATRA J, MITRA A, TYAGI H, et al. Iron oxide nanorods as high-performance magnetic resonance imaging contrast agents[J]. Nanoscale, 2015, 7: 9174.
[308] MITRA A, MOHAPATRA J, MEENA S S, et al. Verwey tran-

sition in ultrasmall-sized octahedral Fe_3O_4 nanoparticles[J]. Journal of Physical Chemistry C, 2014, 118: 19356 - 19362.

[309] CULLEN J R, CALLEN E. Collective electron theory of the metal-semiconductor transition in magnetite[J]. Journal of Applied Physics, 1970, 41: 879 - 880.

[310] BRABERS J H V J, WALZ F, KRONMULLER H. A model for the Verwey transition based on an effective interionic potential[J]. Physica B, 1999, 266: 321 - 331.

[311] WALZ F. The Verwey transition-a topical review [J]. Journal of Physics-Condensed Matter. 2002, 14: R285 - R340.

[312] JIANG Q, LI J C, CHI B Q. Size-dependent cohesive energy of nanocrystals [J]. Chemical Physical Letters, 2002, 366: 551 - 554.

[313] LI J W, YANG L W, ZHOU Z F, et al. Bandgap modulation in ZnO by size, pressure, and temperature [J]. Journal of Physical Chemistry C, 2010, 114: 13370.

[314] OUYANG G, SUN C Q, ZHU W G. Atomistic origin and pressure dependence of band gap variation in semiconductor nanocrystals [J]. Journal of Physical Chemistry C, 2009, 113: 9516.

[315] ZHU Y F, LANG X Y, JIANG Q. The effect of alloying on the bandgap energy of nanoscaled semiconductor alloys[J]. Advanced Functional Materials, 2008, 18: 1422.

[316] SHAN W, WALUKIEWICZ W, AGER III J W, et al. Pressure dependence of the fundamental band-gap energy of CdSe[J]. Applied Physics Letters, 2004, 84: 67.

[317] BIERING S, SCHWERDTFEGER P. A comparative density functional study of the low pressure phases of solid ZnX, CdX, and HgX: Trends and relativistic effects[J]. Journal of Chemical Physics, 2012, 136: 034504.

[318] SOWA H, AHSBAHS H. High-pressure X-ray investigation of zincite ZnO single crystals using diamond anvils with an improved shape[J]. Journal of Applied Crystallography, 2006, 39: 169.

[319] SHAN W, WALUKIEWICZ W, AGER III J W, et al. Pressure-dependent photoluminescence study of ZnO nanowires[J]. Applied Physics Letters, 2005, 86: 153117.

[320] HUSO J, MORRISON J L, HOECK H, et al. Pressure response of the ultraviolet photoluminescence of ZnO and MgZnO nanocrystallites [J]. Applied Physics Letters, 2006, 89: 171909.

[321] IRIMPAN L, NAMPOORI V P N, RADHAKRISHNAN P, et al. Size dependent fluorescence spectroscopy of nanocolloids of ZnO[J]. Journal of Applied Physics, 2007, 102: 063524.

[322] VISWANATHA R, SAPRA S, SATPATI B, et al. Understanding the quantum size effects in ZnO nanocrystals[J]. Journal of Materials Chemistry, 2004, 14: 661.

[323] LIN K F, CHENG H M, HSU H C, et al. Band gap variation of size-controlled ZnO quantum dots synthesized by sol-gel method[J]. Chemical Physics Letters, 2005, 409: 208.

[324] MEULENKAMP E A. Synthesis and growth of ZnO nanoparticles[J]. Journal of Physical Chemistry B, 1998, 102: 5566.

[325] GUO L, YANG S, YANG C, et al. Synthesis and characterization of Poly (vinylpyrrolidone) -modified zinc oxide nanoparticles[J]. Chemistry of Materials, 2000, 12: 2268.

[326] HAASE M, WELLER H, HENGLEIN A. Photochemistry and radiation chemistry of colloidal semiconductors. 23. Electron storage on zinc oxide particles and size quantization[J]. Journal of Physical Chemistry, 1998, 92: 482.

[327] WONG E M, BONEVICH J E, SEARSON P C. Growth

kinetics of nanocrystalline ZnO particles from colloidal suspensions[J]. Journal of Physical Chemistry B, 1998, 102: 7770.
[328] ZENG Z P, GAROUFALIS C S, BASKOUTAS S, et al. Electronic and optical properties of ZnO quantum dots under hydrostatic pressure[J]. Physical Review B, 2013, 87: 125302 (7).
[329] PEDERSEN T G. Quantum size effects in ZnO nanowires[J]. Physica Status Solidi (c), 2005, 2: 4026.
[330] TAN S T, CHEN B J, SUN X W, et al. Blueshift of optical band gap in ZnO thin films grown by metal-organic chemical-vapor deposition [J]. Journal of Applied Physics, 2005, 98: 013505.